HUMAN CAPACITY IN THE

ATTENTION ECONOMY

HUMAN CAPACITY IN THE
ATTENTION
ECONOMY

 AMERICAN PSYCHOLOGICAL ASSOCIATION

Published by
American Psychological Association
750 First Street, NE
Washington, DC 20002
https://www.apa.org

Order Department
https://www.apa.org/pubs/books
order@apa.org

In the U.K., Europe, Africa, and the Middle East, copies may be ordered from Eurospan
https://www.eurospanbookstore.com/apa
info@eurospangroup.com

Typeset in Charter and Interstate by Circle Graphics, Inc., Reisterstown, MD

Printer: Sheridan Books, Chelsea, MI
Cover Designer: Blake Logan, New York, NY

Library of Congress Cataloging-in-Publication Data

Names: Lane, Sean M., editor. | Atchley, Paul, editor.
Title: Human capacity in the attention economy / edited by Sean Lane and
  Paul Atchley.
Description: Washington, DC : American Psychological Association, [2021] |
  Includes bibliographical references and index.
Identifiers: LCCN 2020012941 (print) | LCCN 2020012942 (ebook) |
  ISBN 9781433832079 (paperback) | ISBN 9781433832468 (ebook)
Subjects: LCSH: Cognitive science. | Information technology—Psychological
  aspects. | Information society—Psychological aspects.
Classification: LCC BF311 .H7637 2021 (print) | LCC BF311 (ebook) |
  DDC 153—dc23
LC record available at https://lccn.loc.gov/2020012941
LC ebook record available at https://lccn.loc.gov/2020012942

https://doi.org/10.1037/0000208-000

Printed in the United States of America

10 9 8 7 6 5 4 3 2 1

Contents

Contributors

Paul Atchley, PhD, Department of Psychology, University of South Florida, Tampa, FL, United States

Ruth Ann Atchley, PhD, Department of Psychology, University of South Florida, Tampa, FL, United States

Francesco Biondi, PhD, Faculty of Human Kinetics, University of Windsor, Windsor, Ontario, Canada

Joel M. Cooper, PhD, Department of Psychology, The University of Utah, Salt Lake City, UT, United States

Douglas Getty, BS, Learning Research & Development Center, University of Pittsburgh, Pittsburgh, PA, United States

David N. Greenfield, PhD, MS, The Center for Internet and Technology Addiction, West Hartford, CT; Department of Psychiatry, School of Medicine, University of Connecticut, Farmington, CT; and Greenfield Recovery Center, Leyden, MA, United States

Steven G. Greening, PhD, Cognitive and Brain Sciences, Department of Psychology, Louisiana State University, Baton Rouge, LA, United States

Linda A. Henkel, PhD, Department of Psychology, Fairfield University, Fairfield, CT, United States

Rachel J. Hopman, PhD, Department of Psychology, Northeastern University, Boston, MA, United States

Sean Lane, PhD, College of Arts, Humanities, and Social Sciences, University of Alabama in Huntsville, Hunstsville, AL, United States

Kacie Mennie, PhD, Department of Psychology and Behavioral Sciences, Louisiana Tech University, Ruston, LA, United States

Robert A. Nash, PhD, Department of Psychology, Aston University, Birmingham, England

Justin A. Paton, BS, Department of Psychology, Fairfield University, Fairfield, CT, United States

David L. Strayer, PhD, Department of Psychology, The University of Utah, Salt Lake City, UT, United States

Kevin Yee, PhD, Academy for Teaching and Learning Excellence and Office of Undergraduate Studies, University of South Florida, Tampa, FL, United States

Acknowledgments

We owe a debt of gratitude to the many people who made this project a reality.

We first thank the scientists who contributed to this volume for sharing their expertise, creativity, and insight about the impact of information technology on the human experience. This is a milestone—not an end point—for a conversation that started among us more than a decade ago. We wish them all great success in their future research endeavors and look forward to our continuing discussions.

We also thank Christopher Kelaher, Kristen Knight, and the American Psychological Association Books staff. We deeply appreciate your commitment to this project and your help overcoming the challenges along the way.

Both of us acknowledge our colleagues and institutions for their support during this project. I (SL) appreciate my supportive colleagues in the Department of Psychology at Louisiana State University and, subsequently, at The University of Alabama in Huntsville. Thank you to Kacie Mennie for your help completing this project, to Daniella Cash for your assistance during its early stages, and to the graduate and undergraduate students who worked in my laboratory over the years and never failed to inspire me with their curiosity and commitment. I (PA) am grateful for the support of colleagues in the Department of Psychology at the University of Kansas and, subsequently, at the University of South Florida, Tampa.

Finally, we thank family and friends who were there for us before and during the long process of completing this volume. I (SL) thank my parents, Philip and Mary Jo Lane, for cultivating a desire to understand the world and the intellectual humility to avoid easy answers. In addition, I am deeply grateful to Barbara Basden, Maria Zaragoza, and Marcia Johnson, all of whom played a major role in my development as a scientist. Thank you, Jennifer and Brittney, for your support and encouragement. I (PA) thank my many mentors and collaborators throughout my career, with special acknowledgment to Ruth Ann, my collaborator for life.

HUMAN CAPACITY IN THE

ATTENTION ECONOMY

INTRODUCTION

Defining the Issue and the Structure of This Book

PAUL ATCHLEY AND SEAN LANE

Although it is clear that the internet age has suffused modern life, what exactly is its impact? The popular answer to this question seems to fall into two camps. One is that technology has had an overwhelmingly positive effect on human beings and that life is only going to get better as technology advances (Kurzweil, 2006). The other emphasizes that this influence has primarily been negative, making us less safe and more vulnerable (Carr, 2010; Turkle, 2011b).

Books aimed toward a general audience that examine the impact of the internet age on humans present little or no theory or empirical research. One such book is *Alone Together: Why We Expect More From Technology and Less From Each Other*, by Sherry Turkle (2011a), which argues that in an age in which we have nearly unlimited ability to communicate with each other, we are facing a new type of social isolation. Another example is *The Shallows: How the Internet Is Changing the Way We Think, Read and Remember*, by Nicholas Carr (2010), a *New York Times* best seller, which asks what aspects of our human experience are we sacrificing for new technology. These are excellent books, but they lack an overarching theoretical framework and a grounding in empirical research.

https://doi.org/10.1037/0000208-001
Human Capacity in the Attention Economy, S. Lane and P. Atchley (Editors)
Copyright © 2021 by the American Psychological Association. All rights reserved.

In this volume, we provide a different, more nuanced, and theoretically grounded perspective to the question of the impact of the internet age on human cognition. The basic premise is that information technology (IT) is purposely designed to attract and hold attention, a valuable and limited resource. We must understand how this "attention economy" interacts with attending to technological devices to affect how we think, feel, and behave in both positive and negative ways. We have previously offered a general framework (Atchley & Lane, 2014) to articulate how attentional and cognitive processes are impacted by IT, and how more complex aspects of human experience are affected as a result.

The current volume presents theoretically motivated research on the effects of technology on thinking, feeling, and behavior across a diverse set of research domains. The aim is for the theoretical and empirical research discussed in these chapters to help generate new theory and research on these topics. The contributors of this volume have thought carefully about the implications of their research for the everyday world, and the nuanced view they present is one that is both realistic and optimistic.

We hope this volume will appeal to a broad audience, from researchers to students. Researchers in broad areas including psychology, sociology, design, computer science, and engineering will find provocative ideas in this volume. Within psychology, researchers in human factors and engineering psychology, environmental psychology, applied cognitive psychologists (perception, attention, memory), emotion (particularly those focusing on cognitive control), and clinical psychology (particularly those studying technology addiction) will find topics that intersect with their work. And we hope that this book will appeal to advanced undergraduate and graduate students who are taking classes in areas such as human factors, interface design, computer science, and technology and society, and who, themselves, are developing their own research agenda to answer the critical question of how IT will impact the human experience.

THE ISSUE

The rise of IT in the form of computers, mobile devices, and internet-connected appliances has become commonplace. In this millennium alone, the percentage of adults using the internet in the United States has gone from a slim majority (52%) in 2000 to near totality (89%) in 2018. Moreover, about one quarter of internet users are on the internet "almost constantly," an increase from previous surveys (Perrin & Kumar, 2019). The number is closer to one third for those who access the internet with a mobile device, and nearly two thirds (64%) of American adults own a smartphone (Smith, 2015). And in the young adult demographic (18- to 29-year-olds), the number of those "almost constantly" on the internet according to the same survey (Perrin & Kumar, 2019) is (what some would consider to be an alarming) 39%.

The depth of use has also increased. With respect to mobile access, the term *phone* is almost not an accurate term to describe smart devices. "Traditional" use of the devices (talking) has actually decreased significantly over time (by 25% between 2012 and 2015; Gaskill, 2016). Communication via texting or equivalent applications, video, and new modalities, such as Snapchat, is quickly replacing voice communications. But synchronous communication probably accounts for a diminishing portion of device use. The internet, and all that it brings, and the flexibility of the programs that run on the devices that deliver it allow users to curate their own experience. This permits individuals to build the environment of information as they see fit.

The internet has changed the way we think, feel, and behave. Or, more accurately, it has interacted with the ways in which we are already wired to think, feel, and behave to produce extreme versions of natural tendencies. For example, curiosity is a natural human trait as is sharing what we learn. Curiosity applied to nonvirtual social situations is usually limited by social expectations of propriety and privacy. When the rise of the virtual environment removes those limitations, only unbridled curiosity and the ability to share the results of that curiosity remains. Similarly, the requirement to attend to others in a social situation is supported by the presence of common activities and by the absence of competing stimuli. When we place a device in a person's hands that can deliver a tailor-made, rewarding experience, it competes for time with other activities, including socialization. And we are wired to attend to the opinions and actions of others. But we can now choose to only surround ourselves with like opinions even when many of those "opinions" are artificially augmented by nonhuman actors. A measure of how "the internet has changed everything" is that these new ways of thinking, feeling, and behaving have given rise to a new internet-inspired lexicon.

Doxing, phubbing, and bots are terms associated with these new ways the internet, and the devices associated with them, interact with the ways we are already wired to think, feel, and behave. *Doxing* refers to using the internet to research and broadly share private information. "Nosy" and "gossipy" neighbors have always been a reality, but the internet has changed the concept of "neighbor," adding more neighbors and increasing their anonymity as well as the tools and reach of their "gossip." The power and anonymity of the internet has changed the nature of privacy and added doxing to our lexicon. Now "neighbors" can find out aspects of our lives that we have not knowingly shared and may not wish to be shared. *Phubbing*, or "phone snubbing," refers to focusing on a phone, rather than a person, in a social situation (Chotpitayasunondh & Douglas, 2016). The focus of a social situation is others or the event that drew them together, creating a kind of "shared attention" (Shteynberg, 2015) by definition. But the addition of a personally curated information environment in each hand brings a strong competition to social interaction and adds phubbing to our language. Bots are a more recent internet-inspired phenomenon that many are only recently learning about. *Bots* often refer to automated social media accounts that can mimic specific viewpoints and port content. In some cases, bot content may comprise two thirds of posted content (Wojcik et al., 2018). People have always gravitated to sources of opinion that match their own and have been swayed by what is seemingly the "majority" view. But the ability to choose to follow only like-minded sources and the majority view enhanced by armies of nonhumans leads to increasing polarization.

Whether the changes that the internet will ultimately bring will be beneficial or harmful is one open to debate, and perhaps an answer will have no final form. A recent aggregation of the opinions of thought leaders, researchers, and others found that a majority see promise, but about one third predict more harm than good for well-being in a tech-saturated world (Anderson & Rainie, 2018). And, as of the time of final preparation of this volume, the global COVID-19 pandemic has, in many ways, added a layer of complexity as people worldwide use the internet to replace many of their daily face-to-face interactions. The purpose of this volume is not to form an opinion on the "goodness" or "harmfulness" of the internet and technology. What this volume sets out to accomplish is to develop a way of looking at the impact of both through the lens of human capacity. Specifically, we choose to look at the issue by considering how humans, as limited information processors, interact with an almost infinite informational environment. In other words, the answer to the question about how the internet will affect us lies less with the internet and all that it brings and more with what we bring to it with the cognitive capacity we have now.

OVERVIEW OF THE VOLUME

The purpose of *Human Capacity in the Attention Economy* is to look more closely at how our cognitive systems, unchanged over millennia, are responding to an information environment that is often crafted to take advantage of what we now know about its inner workings. The purpose is not to judge the good or ill of those responses but to provide a framework for understanding them. The book is divided into three parts.

Part I, How Information Technology Influences Behavior and Emotion, examines how technology overcomes our more rational modes of thought. The first chapter in this part, "Digital Distraction: What Makes the Internet and Smartphone So Addictive?" (Chapter 2), focuses on the most extreme case of compulsion: technology addiction. In this chapter, a practitioner with a long history of frontline work on technology addiction has outlined his observations about why some of us take technology use to an unhealthy extreme. The author explores if internet addiction is special or if it is simply a new way to provide a distraction from feelings of isolation and lack of social support. The irony of social media use as a response to social isolation when it leads to increased isolation in some users parallels other substance abuse issues. In the next chapter, "Information Technology and Its Impact on Emotional Well-Being" (Chapter 3), researchers in the field of emotion explore the processes of emotion generation and emotion regulation as impacted by IT. Although there is much to be concerned about regarding how technology impacts emotion regulation, the authors also explore whether there might be a sweet spot for technology to support positive emotional outcomes.

Part II, How Information Technology Influences Cognition and Performance, explores the proposal that technology can improve the cognitive processes underlying learning while it, at the same time, results in decrements to other processes and to performance. The first chapter in this section, "Information Technology and Learning" (Chapter 4), looks at an area in which early technological improvements showed great promise: improving learning. This chapter provides a nice caution from adopting the assumption that everything in an attention economy is unhealthy. But it also demonstrates than an extreme of a good thing can be detrimental. The next chapter, "'Say Cheese!': How Taking and Viewing Photos Can Shape Memory and Cognition" (Chapter 5), makes a similar argument that technology can produce benefits and harms. Furthermore, by examining the effect of taking pictures on memory, this chapter shows how seemingly positive practices that divide attention can produce unintended effects on specific cognitive processes that are not obvious to the person who is engaging in them. The final chapter in the section, "The Multitasking Motorist and the Attention Economy" (Chapter 6), takes on one of the most publicized places in which the attention economy produces costs: driver safety. In this chapter, the authors continue the theme started in the previous chapter about effects of the attention economy on specific cognitive processes and expand it to a wider range of processes necessary for safe driving.

Part III, Getting Away and Looking Forward, ties the themes of the attention economy and imagines "what could be" in the years ahead. The first chapter in this part, "How Nature Helps Replenish Our Depleted Cognitive Reserves and Improves Mood by Increasing Activation of the Brain's Default Mode Network" (Chapter 7), brings us back to the beginning, closer to the time of William James, to understand the role of brain networks that are active when we are not in an attention economy. One way to understand what is lost when attention is depleted is to understand what happens to us when attention is replenished. This chapter explores the idea that "tuning out, turning off, and getting outside" can change us in fundamental and positive ways. In the final chapter of the book, "Charting a Way Forward: Navigating the Attention Economy" (Chapter 8), we provide a summary of all that is positive, thought provoking, and concerning in the exploration of the attention economy in more detail. As the reader will discover, there is much to be gained if we exert control and much to be lost if we allow ourselves to live in, as James (1890) put it, "the confused, dazed, scatterbrained state which in French is called *distraction*" (p. 404).

REFERENCES

Anderson, J., & Rainie, L. (2018, April 17). *The future of well-being in a tech-saturated world*. Pew Research Center. http://www.pewinternet.org/2018/04/17/the-future-of-well-being-in-a-tech-saturated-world/

Atchley, P., & Lane, S. (2014). Cognition in the attention economy. In B. H. Ross (Ed.), *Psychology of learning and motivation* (Vol. 61, pp. 133–177). Academic Press. https://doi.org/10.1016/B978-0-12-800283-4.00004-6

Carr, N. (2010). *The shallows: How the internet is changing the way we think, read and remember*. Atlantic Books.

Chotpitayasunondh, V., & Douglas, K. M. (2016). How "phubbing" becomes the norm: The antecedents and consequences of snubbing via smartphone. *Computers in Human Behavior, 63*, 9–18. https://doi.org/10.1016/j.chb.2016.05.018

Gaskill, B. (2016, July 8). *Smartphone voice calls will decline further in 2016*. The VoIP Report. http://thevoipreport.com/article/smartphone-voice-calls-will-decline-further-in-2016/

James, W. (1890). *The principles of psychology, Vol. 1*. Internet Archive. https://archive.org/details/theprinciplesofp01jameuoft

Kurzweil, R. (2006). Reinventing humanity: The future of machine–human intelligence. *The Futurist, 40*(2), 39–40, 42–46. http://www.singularity.com/KurzweilFuturist.pdf

Perrin, A., & Kumar, M. (2019, July 25). *About three-in-ten U.S. adults say they are "almost constantly" online*. Pew Research Center. https://www.pewresearch.org/fact-tank/2019/07/25/americans-going-online-almost-constantly/

Shteynberg, G. (2015). Shared attention. *Perspectives on Psychological Science, 10*(5), 579–590. https://doi.org/10.1177/1745691615589104

Smith, A. (2015, April 1). *The smartphone difference.* Pew Research Center. http://www.pewinternet.org/2015/04/01/us-smartphone-use-in-2015/

Turkle, S. (2011a). *Alone together: Why we expect more from technology and less from each other.* Basic Books.

Turkle, S. (2011b). *Life on the screen: Identity in the age of the internet.* Simon and Schuster.

Wojcik, S., Messing, S., Smith, A., Rainie, L., & Hitlin, P. (2018, April 9). *Bots in the Twittersphere.* Pew Research Center. http://www.pewinternet.org/2018/04/09/bots-in-the-twittersphere/

1

A GENERAL FRAMEWORK FOR UNDERSTANDING THE IMPACT OF INFORMATION TECHNOLOGY ON HUMAN EXPERIENCE

PAUL ATCHLEY, SEAN LANE, AND KACIE MENNIE

As stated in the Introduction to this volume, the goal of the chapters that follow is to provide a view of the impact of changes brought about by the information economy and the internet age from the standpoint of empirical and theoretically grounded work in human cognition. Work that has been classically referred to as "cognitive science" is spread across a large number of domains, making development of a common theoretical framework a significant challenge. We attempt to meet that challenge by trying to understand the most basic currency of an information economy. The most obvious answer might be taken to be information itself, but information that exists alone and without consideration of a human operator is not as valuable as information that is acted on by a person. (This is true, at least, until automated "intelligent" systems are able to act on information on their own.)

One example of information that has greatest value when it is in the awareness of a human operator can be found in the context of driving. Vehicles today are designed with increasingly sophisticated sensor packages aimed at detecting a variety of aspects of the driving environment. For example, forward-looking cameras and forward-looking radar systems can judge the distance to vehicles in front of a driver. Computations that

https://doi.org/10.1037/0000208-002
Human Capacity in the Attention Economy, S. Lane and P. Atchley (Editors)

measure this information over time reveal changes in distance. This information can be used to alert a driver when the change in distance for a given vehicle speed is rapid enough to suggest a collision might take place. That information is important, but it is only valuable if the driver acts on it. (Unless, of course, the vehicle itself acts on it without driver intervention.) The key here is that the driver must have the ability to pay attention to the information for the information to have value to the driver. If the driver is distracted by a phone, for example, they might fail to process the important information the vehicle is presenting.

Chapter 6, titled "The Multitasking Motorist and the Attention Economy," by David L. Strayer and colleagues, examines this example in much more detail. For the current purposes, the example helps us point to what we see as a key underlying variable that can serve as a theoretical cornerstone: human attention. It is an oversimplification to say, "Without attention, there is nothing in human experience," but many attention researchers would agree this statement is not far from the truth. As research on the phenomenon of inattention blindness so clearly illustrates (Mack, 2003; Mack & Rock, 1998), the selection of stimuli in the environment by attention is crucial for even basic awareness of that information to occur. As originally shown by Neisser and Becklen (1975) and later popularized by Dan Simons in his "invisible gorilla" demonstrations (Simons & Chabris, 1999), even information as obvious as a person (or a gorilla) walking through a basketball game fails to reach awareness if attention is occupied.

The basic role of human attention in the information economy/internet age has not been lost on those seeking to understand how to marry technological innovations and human capacity, lending converging validity to the importance of attention as a theoretical cornerstone. The term "attention economy," which serves as a titular foundation for this volume, was first used in 2001 in a book aimed at a business audience by Thomas Davenport and David Beck. *The Attention Economy: Understanding the New Currency of Business* (Davenport & Beck, 2001) was an attempt to help a generation of professionals who were becoming increasingly frustrated by the inattention associated with the information age. While the information economy had brought incredible new opportunity, it had also created new challenges to attention in the workplace with customers and with consumers. It was certainly not the first work to recognize the importance of human attention in business; marketing and advertising had clearly, if not informally, recognized it for many decades, but it was the first work to treat the issue deeply and with reference to the science of human attention.

As documented in the Introduction, there has been a staggering growth of internet use since the Davenport and Beck volume in 2001. And since then, the ubiquity of internet-enabled smart devices has made threats to limited human attention even more portable. In addition, one could argue that competition for attention has become even more fierce because an entire industry has grown around understanding how to attract and hold consumer attention with smart devices. The metric known as *engagement* is the analytic outcome of an economy in which vying for human attention has become necessary to do business. As described in "This Is How Your Fear and Outrage Are Being Sold for Profit" by Tobias Rose-Stockwell (2017), engagement is the "currency of the attention economy" (graphic after para. 22). Social media companies, app makers, and businesses of all kinds are developing more sophisticated ways to hold the attention of an individual user by tracking and understanding their engagement patterns in real time using automated systems that constantly learn and improve, often with little to no oversight about the accuracy or even humanity underlying the information. (The point about accuracy and humanity makes it a good place to add to the previous list the internet-inspired term *clickbait*, an attention-grabbing item designed to get a user to select a link to internet content.) An understanding of the role of attention is a clear focus for practitioners and is a useful unifying concept for research as well.

WHAT IS "ATTENTION" IN THE ATTENTION ECONOMY?

It is to the concept of attention that we turn to as a theoretical cornerstone on which to build a more theoretically grounded framework to understand human capacity in the information economy/internet age. It is important to note that we are referring to the concept of attention generally. We have previously offered a general framework to articulate how information technology impacts basic attentional and cognitive processes (Atchley & Lane, 2014), and how more complex aspects of human experience are affected as a result. This framework is grounded in classic information processing models of human cognition (Massaro & Cowan, 1993) of the sort used in human factors engineering (Wickens & Hollands, 2000; Wickens et al., 2003). One important aspect of these models is the notion of *attentional resources*, or how much capacity an individual has to process stimuli. Another is the concept of *attentional selection*, which is the process that would lead a driver to select and attend to information from an auditory conversation on a phone at the expense of selection and processing of information from visual stimuli on the roadway. It is important to recognize that capacity changes with practice, and attention changes over time.

We could have attempted to incorporate many aspects of attention, such as models of visual search (e.g., Wolfe, 2010), models of perceptual salience (e.g., Itti & Koch, 2000), aspects of attention capture (Johnston et al., 1990; Yantis & Hillstrom, 1994), or concepts in sustained attention/vigilance (Thomson et al., 2015). However, information processing/human factors models seem appropriate as a generalist approach given attempts to "engineer" the control of human attention previously discussed. This choice is not intended to break any new theoretical ground. Instead, it is meant to focus the discussion of an attention economy on a few key aspects of attention that appear to be the most critical.

To define what we mean by "attention" when we use the term "attention economy," it would be difficult to not start a discussion without reference, as many have done, to how it was described by William James (1890) in his chapter "Attention":

> Everyone knows what attention is. It is taking possession of the mind, in clear and vivid form, of one out of what seem several simultaneously possible objects or trains of thought, localization, concentration, of consciousness are of its essence. It implies withdrawal from some things in order to deal effectively with others, and is a condition which has a real opposite in the confused, dazed, scatter brained state which in French is called *distraction*, and *Zerstreutheit* in German. (pp. 403–404)

James's simple description still resonates today because it describes the phenomenology of attention so well, even if it does not fully describe the cognitive processes that underlie it. Key in his description and subsequent chapter, and in our framing of attention in an attention economy, are a few points. First, there is competition in cognition or "several simultaneously possible objects or trains of thought." In other words, attention can be used in a variety of ways, but it ends up being deployed in only one, or a very small number, of them. Attention is limited. Second, the presence of attention leads to what might be termed *higher order processes*, such as "localization, concentration, of consciousness" (p. 404). Third, in his chapter on attention, James eloquently described that attention is subject to conscious control: "My experience is what I agree to attend to" (p. 402). He also acknowledged that some varieties of attention are "passive, reflex, non-voluntary, effortless" (p. 416). James noted that the process of attention changes over time: "In mature age we have generally selected those stimuli which are connected with one or more so-called permanent interests and our attention has grown irresponsive to the rest" (p. 417). As we note, especially in the chapter on addiction (see Chapter 2, "Digital Distraction: What Makes the Internet and Smartphone So Addictive?"), the ability of attention to change over time has important implications for an attention economy.

The description of James (1890) leads us to consider four general aspects of attention that are useful for the current volume. The four key aspects of the model are the following:

1. Attention is limited.
2. Attention is a key requirement for many cognitive processes.
3. Attention is subject to both internal control and external manipulation.
4. Attentional control is a developmental process, and what attracts attention can change.

ATTENTION IS LIMITED

In his answer to the question, How many things can we attend to at once?, James (1890) wrote,

> If, then, by the original question, how many ideas or things can we attend to at once, be meant how many entirely disconnected systems or processes of conception can go on simultaneously, the answer is *not easily more than one, unless the processes are very habitual; but then two, or even three*, without very much oscillation of the attention. (p. 409; italics added)

The notion of limited attention originates from the basic fact that there are so many things we can possibly perceive, think about, or act on at any one time. Given that we do not perceive, think about, or act on all of them implies some limit to our ability to do so and prompts the simple question of how much we can actually attend to. The notion of limits is also prompted by our own experience when attention is engaged:

> Most people probably fall several times a day into a fit of something like this: The eyes are fixed on vacancy, the sounds of the world melt into confused unity, the attention is dispersed so that the whole body is felt, as it were, at once, and the foreground of consciousness is filled, if by anything, by a sort of solemn sense of surrender to the empty passing of time. (James, 1890, p. 404)

"The abolition of this condition is what we call the awakening of the attention" (p. 405).

The specifics of the limits to attention have been the foundation of research on human attention since the time of James, and what we know now helps us understand some of the subtler ways we process information in the absence of attention. But we are still left with the idea that attention is limited. For the purposes of understanding attention as a limited resource producing an attention economy, we previously focused on the concepts of resources and bottlenecks (Atchley & Lane, 2014). While these are classic concepts that have been refined over time by more recent work, they are useful general descriptions for the purposes of tying together the wide domain of phenomena covered in this volume. The resource view of attention has been criticized as not advancing the progress of attention theory (Navon, 1984) despite a long history of the practice (Humphreys & Revelle, 1984; Kahneman, 1973; Navon & Gopher, 1979; Norman & Bobrow, 1975; Wickens, 1984). However, it works well to describe the phenomenology of attention, and it has merit as a way of understanding what we are beginning to more fully grasp about the neurophysiology of thought arising from research using a suite of neuroscience techniques. To tie in with the themes of Chapter 6, even early work on the distribution of metabolic resources clearly demonstrates that attempts to coordinate multiple "disconnected systems or processes of conception," as James (1890, p. 409) would call them (in this case, using social technologies while driving), shows that the metabolic resources to simultaneously activate them are lacking (Cohen et al., 1988; Just et al., 2008; Kinomura et al., 1996; Lewin et al., 1996; Pardo et al., 1991; Shomstein & Yantis, 2006).

The bottleneck explanation of reduced performance has a long history grounded in empirical work, such as the early studies of Telford (1931) on the refractory period. (The reader is encouraged to read Welford [1952] for a more thorough overview of early work.) The concept of processing bottlenecks is a useful way of considering exactly how limits start to influence specific cognitive processes. The bottleneck approach can be viewed as a more specific version of resource theory: Attention is limited by restrictions on the capacity of specific cognitive processes required to perform most complex cognitive functions. For example, to preview Chapter 5 ("Say Cheese!": How Taking and Viewing Photos Can Shape Memory and Cognition), the process of encoding information is limited, and therefore memory representations can be subject to competing priorities at the point of the perceptual experience. Temporally farther downstream in cognitive processing from encoding, the long history of dual-task and psychological refractory period research have identified significant bottlenecks in response selection and response execution. While remaining agnostic with respect to the number and nature of processing bottlenecks, it is important to acknowledge that the concept of information processing bottlenecks warns us that limiting attention can produce multiple effects ranging from a failure to encode information to a failure to act.

ATTENTION IS A KEY REQUIREMENT FOR MANY COGNITIVE PROCESSES

The bold statement "attention is everything" would carry the core concept of attention too far, but it is arguably not too distant from the truth. Examining where this conceptual framing is incorrect helps to understand what we are defining as attention, what the value of attention is, and how living in an attention economy shapes us. In the framework we have previously suggested, we noted the distinction between automatic and controlled processing can provide a useful dichotomy for understanding the role attention plays in specific cognitive processes. Again, James (1890) captured this idea by stating attention can either be "passive, reflex, non-voluntary, effortless" or "active and voluntary" (p. 416). In the classic version of automatic versus controlled process (Posner & Snyder, 1975; Schneider & Shiffrin, 1977), *automatic processes* are those that James described as "habitual" and, as he alluded, can be performed simultaneously with other tasks. Automatic processes are triggered by environmental context rather than conscious intent. *Controlled processes* are those that require more conscious intent or attention to complete. Controlled processes, such as driving using a manual transmission, can become automatic with training. Although this simple dichotomy has been criticized on a number of grounds (e.g., Hirst et al., 1980; Logan, 1988; Pashler, 1998), including that it is more continuum than dichotomy, it is still a useful framing device for the current work.

In the model of an attention economy, we propose that understanding when the resources required for controlled process begin to shift other processes into more automatic modes points out an important distinction between what we might call inattention and impoverished attention. Much of the criticism of technology, for example, in the field of driving safety, has focused on the idea that tasks competing for attention, such as phone activity while driving, produce states of inattention that lead to failure to perceive the environment at an even basic level. But much of the real risk probably falls into the realm of *impoverished attention*, in which attention is not burdened enough for someone to completely lose track of other tasks, but the reduction in attention changes the way that a task is performed. In the safety example, monitoring a phone conversation reduces visual scanning (Nunes & Recarte, 2002; Recarte & Nunes, 2003), limits neurophysiological responses to items the eyes fixate on (Strayer & Drews, 2007), and slows perceptual processing and reaction time (Atchley & Dressel, 2004). In many cases, attention is not absent, causing drivers to fail to maintain lane

position and headway; rather, it is just functionally impaired or impoverished to the degree it cannot support timely reactions to unexpected events.
As we see in Chapter 7 ("How Nature Helps Replenish Our Depleted Cognitive Reserves and Improves Mood by Increasing Activation of the Brain's
Default Mode Network"), reduces safety is only one risk associated with a
shift to more automatic processes. The attention economy can deplete other
abilities, like creativity, impoverishing us in even more subtle ways.

ATTENTION IS SUBJECT TO BOTH INTERNAL CONTROL
AND EXTERNAL MANIPULATION

When technology draws our focus and shifts what should be controlled
processes to more automatic, cue-driven ones, the loss of conscious control
has important implications for our ability to direct of our own choices.
Kahneman's System 1 and System 2 approach (Kahneman, 2011) is a useful
framework for thinking about how attention can be subject to both internal
control and external manipulation in an attention economy. In his framework, decisions can be influenced by either by relatively automatic processes
that are fast but subject to less voluntary control (*System 1*), or by slower,
more cognitively demanding processes that are subject to conscious inspec-
tion (*System 2*; Kahneman, 2011). Hijacking System 2 by capturing and
holding conscious attention can lead to a shift to these more automatic
System 1 processes for other tasks. Furthermore, portals to the internet,
including web pages and smart devices, are designed in a way that promotes
a constant capture of System 2 resources. There is always another link or
another app vying for attention. A constant shifting in attention between
stimuli results in a depletion of System 2 resources necessary for the exercise of willpower, further shifting us toward System 1 thinking (Baumeister
et al., 1998).

The concepts of emotion and emotional regulation are core to the
notion of attention in the attention economy for a few reasons. First, again
using Kahneman's (2011) framework, the fast System 1 is often aligned to
more emotional components of decision making. System 2, which requires
attention of the "active and voluntary" sort described by James (1890), is
tasked with more slowly evaluating possible decisions suggested by first,
more emotional, reactions. The implications of this is that an attention
impoverished by a resource deficit is more prone to become a system
that is responding in an emotional rather than a rational way. The internet
is designed to change the way we think and feel by understanding individual "likes and dislikes" to target and enhance the emotional strength

of messages so they amplify engagement or the currency of the attention economy (Rose-Stockwell, 2017). As the internet captures more of attention using emotionally tuned cues, it in turn leaves us with fewer conscious resources to evaluate our own moment-to-moment decision making. As outlined in Chapter 3 ("Information Technology and Its Impact on Emotional Well-Being"), this is only one of myriad ways in which technology and the attention economy it creates can affect emotions and emotion processing.

ATTENTIONAL CONTROL IS A DEVELOPMENTAL PROCESS, AND WHAT ATTRACTS ATTENTION CAN CHANGE

When it comes to the development of attention, William James (1890) again described attention in a way that seems to anticipate how technology, multimedia, and the internet might affect child development:

> Sensitiveness to immediately exciting sensorial stimuli characterizes the attention of childhood and youth. In mature age we have generally selected those stimuli which are connected with one or more so-called permanent interests, and our attention has grown irresponsive to the rest. But childhood is characterized by great active energy, and has few organized interests by which to meet new impressions and decide if they are worthy of notice or not, and the consequence is that the extreme mobility of the attention with which we are all familiar in children, and which makes their first lessons such rough affairs. (p. 417)

James noted both that attention and what attracts it will change over time. Furthermore, he perceived the risks of youths interacting in an attention economy who lack the tools to adequately decide if the "new impressions" are "worthy of notice or not."

A key component of the model of attention in an attention economy is that attention changes. The control of attention is a developmental process, and what attracts attention is dynamic and malleable. As Uncapher et al. (2017) noted in their review, many results show increased multitasking changing developmental trajectories for the worse and some showing improved development as well as null findings. The question of whether or not the impacts of attention economy on development are positive or negative is less important to the current volume than understanding how an attention economy interacts with the intersection of development of attention processes. As we have reviewed, attention is subject to both internal control and external manipulation. In many ways, much of the design of the internet and the attention economy is about manipulating attention by reducing the influence of internal control mechanisms and creating so-called permanent interests as James (1890) put it. Among teens who use a smartphone, for example, the average time per day of use is more than 4 hours (Common Sense Media, 2015). That certainly appears to be the development of a new so-called permanent interest. Development of this new attentional focus is supported by at least two factors. First, smartphones deliver types of inter-actions (social media) that our brains crave. Social media is known to active reward networks similar much as other, stronger, instrumental rewards do (Tamir & Mitchell, 2012). Talking about ourselves, indicating what we like and dislike, and being socially connected to massive networks that can pro-vide immediate positive feedback are about as attractive to a young brain as anything else. This could be overcome by self-regulation. But, secondly, the ability to regulate overuse of a smartphone is impaired by an incomplete development of self-regulation in teens or what Kahneman (2011) called System 2 thinking.

A brain less capable of regulating the lure of instrumental rewards has implications for development across a wide range of domains (see Uncapher et al., 2017, for a recent review). These include, but are not limited to, emotional development, such as the development of empathy (see Chap-ter 3); safety when attention to a primary task undergoing skill development, such as driving, competes with other stimuli (see Atchley & Strayer, 2017, and Chapter 6, this volume); psychosocial development, such as the ability to exercise willpower; and the structural development of the brain and use of brain networks (see Chapter 7). Rich technology environments also have impacts, both positive and negative, on learning, which Chapter 4 ("Informa-tion Technology and Learning") examines in more detail. In the extreme, the malleability of what attracts and holds our attention can be so imbalanced that it takes on characteristics of addiction (see Chapter 2, "Digital Distraction: What Makes the Internet and Smartphone So Addictive?")

SUMMARY

We are almost a century and a half since William James (1890) wrote about attention in his book *The Principles of Psychology*, yet much of what he wrote seems incredibly relevant today when we try to understand the implica-tions of the attention economy on how we think, feel and behave. Rather than postulate that psychological science has not made much progress since 1890, there is a much simpler explanation: We, and our brains, have not changed much.

Attention is a valuable and limited resource. The processes that support how we think, feel, and behave have bottlenecks and regulatory systems that are impacted by limitations on attention. Attention can be subject to self-control but also can be influenced by external forces that subvert that control. And how we use our limited attention changes with the long-term exposure to those influences as well as exposure occurring at a vulnerable age. The effects are often noticeable, but many of them are not obvious until we examine them in more detail. That examination is the focus of the rest of this book.

As the reader examines the remaining work in this volume, they are advised that although attention is the underlying construct in the attention economy, each team of authors approaches the issue from a theoretical framework germane to their own field. Even so, it is important for the reader to be cognizant of how attention plays a role. In some cases, the role of attention is made clear because research—which the authors highlight—has already been done in that direction. In other cases, the role of attention has not been formally realized. When considering the role of an attention economy, consider how the internet changes how we think, feel, and behave. We offer a few thoughtful questions to keep in mind. How is a process that relies on a limited resource affected when attention is made more limited by a source designed to attract it? What happens when more practiced aspects of a task remain intact, but more effortful components are negatively impacted by distraction? Is there evidence that capacity can be increased through practice or that technology might serve to "enhance" normal performance? How much cognitive control does an individual have, and is there reason to suspect that might change with age, experience, or both? There are undoubtedly other questions, and we hope the reader views each chapter as a solicitation for new questions and possibly new research ideas. After all, at this point, there are far more questions than answers in the attention economy!

REFERENCES

Atchley, P., & Dressel, J. (2004). Conversation limits the functional field of view. *Human Factors, 46*(4), 664–673. https://doi.org/10.1518/hfes.46.4.664.56808

Atchley, P., & Lane, S. (2014). Cognition in the attention economy. In B. H. Ross (Ed.), *Psychology of learning and motivation* (Vol. 61, pp. 133–177). Academic Press. https://doi.org/10.1016/B978-0-12-800283-4.00004-6

Atchley, P., & Strayer, D. L. (2017). Small screen use and driving safety. *Pediatrics, 140*(Suppl. 2), S107–S111. https://doi.org/10.1542/peds.2016-1758M

Baumeister, R. F., Bratslavsky, E., Muraven, M., & Tice, D. M. (1998). Ego depletion: Is the active self a limited resource? *Journal of Personality and Social Psychology, 74*(5), 1252–1265. https://doi.org/10.1037/0022-3514.74.5.1252

Cohen, R. M., Semple, W. E., Gross, M., Holcomb, H. H., Dowling, M. S., & Nordahl, T. E. (1988). Functional localization of sustained attention: A comparison to sensory stimulation in the absence of instruction. *Neuropsychiatry, Neuropsychology, & Behavioral Neurology, 1*(1), 3–20.

Common Sense Media. (2015). *Fact sheet: Teens and smartphones—The Common Sense census: Media use by tweens and teens.* Common Sense. https://www.commonsense media.org/sites/default/files/uploads/pdfs/census_factsheet_teensandsmart phones.pdf

Davenport, T. H., & Beck, J. C. (2001). *The attention economy: Understanding the new currency of business.* Harvard Business School Press.

Hirst, W., Spelke, E. S., Reaves, C. C., Caharack, G., & Neisser, U. (1980). Dividing attention without alternation or automaticity. *Journal of Experimental Psychology: General, 109*(1), 98–117. https://doi.org/10.1037/0096-3445.109.1.98

Humphreys, M. S., & Revelle, W. (1984). Personality, motivation, and performance: A theory of the relationship between individual differences and information processing. *Psychological Review, 91*(2), 153–184. https://doi.org/10.1037/0033-295X.91.2.153

Itti, L., & Koch, C. (2000). A saliency-based search mechanism for overt and covert shifts of visual attention. *Vision Research, 40*(10–12), 1489–1506. https://doi.org/10.1016/S0042-6989(99)00163-7

James, W. (1890). *The principles of psychology, Vol. 1.* Internet Archive. https://archive.org/details/theprinciplesofp01jameuoft

Johnston, W. A., Hawley, K. J., Plewe, S. H., Elliott, J. M., & DeWitt, M. J. (1990). Attention capture by novel stimuli. *Journal of Experimental Psychology: General, 119*(4), 397–411. https://doi.org/10.1037/0096-3445.119.4.397

Just, M. A., Keller, T. A., & Cynkar, J. (2008). A decrease in brain activation associated with driving when listening to someone speak. *Brain Research, 1205,* 70–80. https://doi.org/10.1016/j.brainres.2007.12.075

Kahneman, D. (1973). *Attention and effort.* Prentice-Hall.

Kahneman, D. (2011). *Thinking, fast and slow.* Macmillan.

Kinomura, S., Larsson, J., Gulyás, B., & Roland, P. E. (1996). Activation by attention of the human reticular formation and thalamic intralaminar nuclei. *Science, 271*(5248), 512–515. https://doi.org/10.1126/science.271.5248.512

Lewin, J. S., Friedman, L., Wu, D., Miller, D. A., Thompson, L. A., Klein, S. K., Wise, A. L., Hedera, P., Buckley, P., Meltzer, H., Friedland, R. P., & Duerk, J. L. (1996). Cortical localization of human sustained attention: Detection with functional MRI using a visual vigilance paradigm. *Journal of Computer Assisted Tomography, 20*(5), 695–701. https://doi.org/10.1097/00004728-199609000-00002

Logan, G. D. (1988). Toward an instance theory of automatization. *Psychological Review, 95*(4), 492–527. https://doi.org/10.1037/0033-295X.95.4.492

Mack, A. (2003). Inattentional blindness: Looking without seeing. *Current Directions in Psychological Science, 12*(5), 180–184. https://doi.org/10.1111/1467-8721.01256

Mack, A., & Rock, I. (1998). *Inattentional blindness.* MIT Press. https://doi.org/10.7551/mitpress/3707.001.0001

Massaro, D. W., & Cowan, N. (1993). Information processing models: Microscopes of the mind. *Annual Review of Psychology, 44*(1), 383–425. https://doi.org/10.1146/annurev.ps.44.020193.002123

Navon, D. (1984). Resources—A theoretical soup stone? *Psychological Review, 91*(2), 216–234. https://doi.org/10.1037/0033-295X.91.2.216

Navon, D., & Gopher, D. (1979). On the economy of the human-processing system. *Psychological Review, 86*(3), 214–255. https://doi.org/10.1037/0033-295X.86.3.214

Neisser, U., & Becklen, R. (1975). Selective looking: Attending to visually specified events. *Cognitive Psychology, 7*(4), 480–494. https://doi.org/10.1016/0010-0285(75)90019-5

Norman, D. A., & Bobrow, D. G. (1975). On data-limited and resource-limited processes. *Cognitive Psychology, 7*(1), 44–64. https://doi.org/10.1016/0010-0285(75)90004-3

Nunes, L., & Recarte, M. A. (2002). Cognitive demands of hands-free-phone conversation while driving. *Transportation Research Part F: Traffic Psychology and Behaviour, 5*(2), 133–144. https://doi.org/10.1016/S1369-8478(02)00012-8

Pardo, J. V., Fox, P. T., & Raichle, M. E. (1991). Localization of a human system for sustained attention by positron emission tomography. *Nature, 349*(6304), 61–64. https://doi.org/10.1038/349061a0

Pashler, H. (1998). *The psychology of attention.* MIT Press.

Posner, M. I., & Snyder, C. R. R. (1975). Facilitation and inhibition in the processing of signals. In P. M. A. Rabitt & S. Dornic (Eds.), *Attention and performance V* (pp. 669–682). Academic Press.

Recarte, M. A., & Nunes, L. M. (2003). Mental workload while driving: Effects on visual search, discrimination, and decision making. *Journal of Experimental Psychology: Applied, 9*(2), 119–137. https://doi.org/10.1037/1076-898X.9.2.119

Rose-Stockwell, T. (2017, July 14). This is how your fear and outrage are being sold for profit. *Medium Media.* https://medium.com/@tobiasrose/the-enemy-in-our-feeds-e86511488de

Schneider, W., & Shiffrin, R. M. (1977). Controlled and automatic human information processing: I. Detection, search, and attention. *Psychological Review, 84*(1), 1–66. https://doi.org/10.1037/0033-295X.84.1.1

Shomstein, S., & Yantis, S. (2006). Parietal cortex mediates voluntary control of spatial and nonspatial auditory attention. *Journal of Neuroscience, 26*(2), 435–439. https://doi.org/10.1523/JNEUROSCI.4408-05.2006

Simons, D. J., & Chabris, C. F. (1999). Gorillas in our midst: Sustained inattentional blindness for dynamic events. *Perception, 28*(9), 1059–1074. https://doi.org/10.1068/p281059

Strayer, D. L., & Drews, F. A. (2007). Cell-phone–induced driver distraction. *Current Directions in Psychological Science, 16*(3), 128–131. https://doi.org/10.1111/j.1467-8721.2007.00489.x

Tamir, D. I., & Mitchell, J. P. (2012). Disclosing information about the self is intrinsically rewarding. *Proceedings of the National Academy of Sciences, 109*(21), 8038–8043. https://doi.org/10.1073/pnas.1202129109

Telford, C. W. (1931). The refractory phase of voluntary and associative responses. *Journal of Experimental Psychology, 14*(1), 1–36. https://doi.org/10.1037/h0073262

Thomson, D. R., Besner, D., & Smilek, D. (2015). A resource-control account of sustained attention: Evidence from mind-wandering and vigilance paradigms. *Perspectives on Psychological Science, 10*(1), 82–96. https://doi.org/10.1177/1745691614556681

Uncapher, M. R., Lin, L., Rosen, L. D., Kirkorian, H. L., Baron, N. S., Bailey, K., Cantor, J., Strayer, D. L., Parsons, T. D., & Wagner, A. D. (2017). Media multitasking and cognitive, psychological, neural, and learning differences. *Pediatrics, 140*(Suppl. 2), S62–S66. https://doi.org/10.1542/peds.2016-1758D

Welford, A. T. (1952). The "psychological refractory period" and the timing of high-speed performance—A review and a theory. *British Journal of Psychology. General Section, 43*(1), 2–19. https://doi.org/10.1111/j.2044-8295.1952.tb00322.x

Wickens, C. D. (1984). Processing resources in attention. In R. Parasuraman & D. R. Davies (Eds.), *Varieties of attention* (pp. 63–102). Academic Press.

Wickens, C. D., Goh, J., Helleberg, J., Horrey, W. J., & Talleur, D. A. (2003). Attentional models of multitask pilot performance using advanced display technology. *Human Factors, 45*(3), 360–380. https://doi.org/10.1518/hfes.45.3.360.27250

Wickens, C. D., & Hollands, J. G. (2000). Complex systems, process control, and automation. In C. D. Wickens & J. G. Hollands (Eds.), *Engineering psychology and human performance* (3rd ed., pp. 538–550). Prentice Hall.

Wolfe, J. M. (2010). Visual search. *Current Biology, 20*(8), R346–R349. https://doi.org/10.1016/j.cub.2010.02.016

Yantis, S., & Hillstrom, A. P. (1994). Stimulus-driven attentional capture: Evidence from equiluminant visual objects. *Journal of Experimental Psychology: Human Perception and Performance, 20*(1), 95–107. https://doi.org/10.1037/0096-1523.20.1.95

HOW INFORMATION TECHNOLOGY INFLUENCES BEHAVIOR AND EMOTION

2 DIGITAL DISTRACTION

*What Makes the Internet and Smartphone
So Addictive?*

DAVID N. GREENFIELD

Almost every day, we hear or read about someone being addicted to the internet. Everywhere we turn, we see people with their faces buried in their smartphones. Why have the internet and the portals through which it is consumed, such as the smartphone, become so compelling and, in some cases, addictive? The basis for my perspective has been shaped by having treated hundreds of patients over 23 years for problems related to their internet and screen use, from which they essentially have suffered from an induced attention disorder. The focus of this chapter is to examine why the internet, regardless of the portal used, is so compelling. I use the terms *internet use disorder* and *internet addiction* interchangeably throughout this chapter.

Time and attention are the functional currency of how we live our lives. Addiction, and more specifically in this case, internet use disorder, facilitates much of its dysfunctional impact by usurping our time and attention to the extent that our lives become so unbalanced that we experience negative consequences. On a practical level, the etiology of this behavior is directly related to how our attentional capacity is highjacked through activation of mesolimbic reward circuits in the brain. The power of all addictions,

https://doi.org/10.1037/0000208-003
Human Capacity in the Attention Economy, S. Lane and P. Atchley (Editors)

especially internet use disorder, is the psychoactive ability to alter mood and consciousness, distort time, and therefore impact healthy and balanced living (Bergmark et al., 2011).

My interest in digital distraction and the addictive power of the internet began in 1997, when I obtained my first good computer and connected to the then lightning-fast, 28.8 megabytes-per-second, dial-up modem. I recall spending many hours in a near-dissociated trance, looking at useless information and rudimentary websites.

Even in these early Wild West days of the internet, with little content online, I found the lure of the internet to produce an altered mood state compelling. The time distorting and mood-elevating power of the internet appeared psychoactive to me, and my early experiences with what I later labeled "the world's largest slot machine" ignited a clinical curiosity into how and why the internet appeared to be a powerful digital drug.

As I was experiencing my own digital distraction, the late Kimberly Young (1998) published a pilot study examining the similarities between gambling addiction and internet behavior. The comparison seemed noteworthy, and it prompted me to partner with ABC News from 1997 to 1998 to conduct one of the first large-scale surveys on internet use behavior (Greenfield, 1999a, 1999b).

Our preliminary data, based on more than 17,000 responses to our virtual addiction survey and using our modified rating system based on pathological gambling criteria in the fifth edition of the *Diagnostic and Statistical Manual of Mental Disorders* (*DSM-5*; American Psychiatric Association, 2013), demonstrated a 5.9% rate of addictive use of the internet (Greenfield, 1999b). Our results, which later became the basis for my book *Virtual Addiction* (Greenfield, 1999c), were presented at the 1999 meeting of the American Psychological Association in Boston, Massachusetts. The Associated Press attended the meeting, and the results of that interview sparked an inextinguishable interest in this topic that has continued to this day. My findings hit a nerve; this new and exciting communication medium called the internet had addictive properties, and this interview reshaped my career in addiction medicine as a specialist in the field of internet addiction.

Our study (Greenfield, 1999b) uncovered several major variables that seemed to be relevant in accounting for the compelling nature of the internet: ease of access, accessibility, perceived anonymity, time distortion, and disinhibition. These variables all contributed to the powerful effect that I labeled *synergistic amplification*, in which the internet modality amplifies the intoxicating impact of stimulating content via the efficient delivery mechanism into our nervous system. We were witnessing the birth of *info-tainment*: pornography, shopping, social content, information, and video gaming readily accessible all in one place.

The field of internet addiction research was virtually nonexistent in the late 1990s when I began my work with the world's latest addiction. It wasn't until many years later, when I had expanded my theories on the addictive nature of the internet by looking at the neurobiology of addiction, that the world caught up with this burgeoning field of concern (Greenfield, 2018; Kuss & Griffiths, 2012; Sohn et al., 2019).

ADDICTION DEFINED

All addictions take time and attention and involve a hyperfocusing on the pleasurable substance or behavior. According to the American Society of Addiction Medicine, addiction is a primary disruption of mood, behavior, and functioning, and, along with compulsive cravings, can lead to negative life impact in the major spheres of living (Ries et al., 2014). The American Society of Addiction Medicine perhaps offers the most succinct definition of addiction, capturing the behavioral components of addiction while simultaneously addressing the neurobiological disruption of the mesolimbic reward circuitry of the brain (Ries et al., 2014; Robbins & Clark, 2015):

> Addiction is a primary, chronic disease of brain reward, motivation, memory and related circuitry. Dysfunction in these circuits leads to characteristic biological, psychological, social and spiritual manifestations. This is reflected in an individual pathologically pursuing reward and/or relief by substance use and other behaviors.

Addiction is characterized by inability to consistently abstain, impairment in behavioural control, craving, diminished recognition of significant problems with one's behaviours and interpersonal relationships, and a dysfunctional emotional response. Like other chronic diseases, addiction often involves cycles of relapse and remission. Without treatment or engagement in recovery activities, addiction is progressive and can result in disability or premature death. ("ASAM Releases New Definition of Addiction," 2011, p. 1)

Although premature death is atypical and infrequent with internet addiction, there are numerous psychological, behavioral, and physiological consequences of protracted internet and video game use. Numerous anecdotal and clinical reports point to various physiologic sequelae, including elevated cortisol, hypertension, deep vein thrombosis, and electrolyte imbalances leading to cardiac dysrhythmias; in addition, obesity, sleep, and metabolic disorders secondary to sedentary behavior have been widely noted (Bibbey et al., 2015; Lemieux & al'Absi, 2016; Reed et al., 2015).

Some controversy in addiction medicine exists over the similarities and differences between substance-based and behavioral (process) addictions (Liese et al., 2018); however, the American Society of Addiction Medicine ("ASAM Releases New Definition of Addiction," 2011) definition of addiction substantively captures the complex biopsychosocial interplay that contributes to addiction as a complex brain-behavior disorder. It appears that behavior alters brain chemistry, neurobiology, and brain structure and the reverse is also true: Biological changes further alter behavior in a recursive interplay.

ADDICTION: A BIOPSYCHOSOCIAL PHENOMENON

Addictions to substances or behaviors all likely involve similar neurobiological and behavioral processes (Jorgenson et al., 2016). It has been well established that the reward structures in the mesolimbic system account for attraction, desirability, and behavioral repetition (learned patterning) of pleasurable behaviors (Grant et al., 2010). Brain circuits involved in the complex biobehavioral phenomenon of addiction include the ventral tegmental area and substantia nigra, amygdala, anterior cingulate, prefrontal cortex, and, most notably, the nucleus accumbens. These same complex circuits also seem to be implicated in internet and screen addiction (S. H. Kim et al., 2011; Kuss & Lopez-Fernandez, 2016) and likely account for the power and potency of how variable reinforcement creates a *maybe factor*—an addictive brain reward pattern in which the anticipation of seeing something we like approximates being glued to a slot machine; this is not only in expectation of a win ("Smartphone Addiction," n.d.) but supports a compulsive pattern of being online. Just like in gambling, the addict chases the possibility of winning, but, in reality, it's being in the game that elevates dopamine even more. This maybe factor is exceptionally compelling when we connect to the internet.

It follows that any substance or behavior that elevates dopamine, impacts cortisol, and modulates mood may possess addictive properties. Furthermore, the anticipation of finding a desirable piece of content online elevates dopamine at twice the level of experiencing desirable content (Blum et al., 2018; Knutson et al., 2001; Schott et al., 2008). This anticipated pleasure increases the likelihood and potential for continued and compulsive use, and neuroimaging studies support similar neurobiological processes for internet addiction as seen in other substance and behavioral addictions.

(Brand et al., 2014; Byun et al., 2009; Jović & Đinđić, 2011; S. H. Kim et al., 2011; Koob & Volkow, 2016; Kuss & Griffiths, 2012; National Institute on Drug Abuse, 2018).

Research, clinical, and anecdotal data on addiction (both behavioral and substance-based) in the United States strongly suggest that social isolation, disconnection, and lack of social supports are predictive of the need to bond (use and abuse) a drug or of intoxicating behavior more than anything else (Adams, 2016; Alexander, 2012; Hari, 2015a; Twenge et al., 2018). Despite genetic factors, which may account for up to 50% of the epidemiologic variance for addiction (Comings, 2006; Goudriaan et al., 2004), it seems that epigenetic and sociobehavioral factors are a significant epidemiological factor in all addictions. In the case of internet addiction, social isolation appears to be a strong contributing factor to the deleterious impact of the internet.

THE EVOLUTIONARY BIOLOGY OF ADDICTION

There may be an evolutionary and adaptive basis for addictive behavior based on the development of mammalian subcortical reward circuits. Reproducing and eating are both appetitive, limbic-based behaviors; are tied to dopamine (pleasure) pathways; and are survival ensuring. We are wired for pleasure seeking and pain avoidance, both of which are biologically adaptive for survival. And these behaviors inadvertently facilitate the addiction potential of other behaviors that piggyback on these primitive reward circuits (Bozarth, 1994; Nesse, 2002; Saah, 2005). The important perspective in understanding internet and smartphone addiction is that developing an addictive behavior pattern with high-pleasure activities is by no means new; ample evidence shows that addiction is as old as early human and mammalian brain development (Saah, 2005) and that mammalian limbic structures are responsible for addictogenic behavior. *Addictogenic behaviors* mimic the simulative capacity or pleasure potential of pleasure behaviors that are linked to adaptive survival. Addiction is not simply the physiologic response of tolerance and withdrawal that we associate with drug and alcohol use disorders and addiction but rather a complex set of biopsychosocial learning patterns and reward circuit conditioning that produces the complex phenomenon of addiction.

The neurohormonal pathways from midbrain to frontal cortex are essentially bypassed during high arousal of hypothalamic-pituitary-adrenal axis activation. This, too, has an adaptive context in that the slower cortical processing may be counterproductive in terms of flight–fight reactivity. It seems there perhaps is an adaptive context to the process of *hypofrontality*—in which an efficient deployment of reward circuit activation leads to a decrease in orbitofrontal processing and a reduction in executive functioning (Ranganath & Jacob, 2016)—and the bypassing of slower processing via frontal cortical executive circuits helps ensure survival via more rapid stimulus–response limbic processing (Greenfield, 2016). We typically see hypofrontality and resultant executive dysfunction in many addictions (Crews & Nixon, 2009). Executive functions involve abstraction, learning and motivation, organizing, planning, attention to tasks at hand, and inhibition of impulsive behaviors. "Impulsiveness generally refers to premature, unduly risky, poorly conceived actions. Dysfunctional impulsivity includes deficits in attention, lack of reflection and/or insensitivity to consequences, all of which occur in addiction (de Wit, 2009; Evenden, 1999)" (Crews & Boettiger, 2009, p. 237).

DIGITAL DRUGS AND THE SLOT MACHINE

The internet, and especially the smartphone, hampers our ability to manage time, energy, and attention, and it can be addictive (Fu et al., 2018; Greenfield, 2015). The internet medium might be described as the world's largest slot machine in that the internet operates on a *variable ratio reinforcement schedule* of neurobiological reinforcement; that is, when you go online for anything, you never quite know what you are going to find, when you are going to find it, or how desirable or pleasurable the content will be. That is how a slot machine works; it is the unpredictability—the maybe factor—that keeps our brains dopaminergically tuned in and attentive to receiving a reward in whatever digital form we find pleasurable. Dopamine is strongly implicated in reward, compulsion, and addiction circuits in the brain (Grant et al., 2010; Moreira & Dalley, 2015). Because the reward is variable and unpredictable, it is therefore highly resistant to extinction (Ferster & Skinner, 1957); we endlessly keep going online because every so often we will find something desirable. Internet service providers and content developers know these behavioral principles well and use this brain science to capture our attention and keep our eyes on our screens. The content is irrelevant; it is about what is salient and impactful—and therefore, potentially enervating—to you.

The smartphone is fast becoming the dominant internet access portal in the world with almost 53% of internet searches derived from mobile devices ("Percentage of Mobile Device Website Traffic Worldwide From 1st Quarter 2015 to 4th Quarter 2019," n.d.). With notifications, we are constantly receiving information pushed through to us as beeps, buzzes, and flashes that tell us something (perhaps pleasurable) is waiting for us to see, and it is that anticipation of receiving possible desirable content that provides the greatest elevation of dopamine. It is this dopamine hit that keeps us pushing the handle or button on a slot machine. We check our phone because *maybe* what we will find will be pleasurable, and afterward, we might get a secondary reinforcing dopamine hit if it *is* something we like. Our smartphone then is like the world's smallest slot machine carried in our pocket or purse, thus significantly increasing its potential for addiction and distractibility.

REWARD DEFICIENCY SYNDROME

Addictions are, in part, maintained by *reward deficiency syndrome* in which normal life seems boring and less potent when compared with highly stimulatory and addictive online content and behavior. This desensitization (dopamine postsynaptic receptor upregulation) involves a weakening of circuits related to naturalistic rewards (e.g., food, sex, social activities, work or academics, delayed gratification). Once a compulsive level of internet use begins, previously reinforcing behaviors decrease in salience, and an increasing amount of time and attention is devoted to internet use.

The end effect is an overall reduction of endogenous levels of accessible dopamine due to progressive downregulation at the dopamine receptor. What I and other practitioners in our clinic see is the reduced ability to derive reward from previously pleasurable experiences, which leaves the patient with an inability to derive and delay gratification for typical life tasks, such as work and academics, as well as self-care and social relationships. Another feature seen in addiction is hypofrontality that possibly is due to evolutionary survival mechanisms related to pleasurable and rewarding experiences derived from food and sex.

Historical analysis of addiction in the United States (Alexander, 2012; Hari, 2015a) suggests that social and community connection, as well as social supports, are predictive factors in addiction. We are hardwired for social connection, and if we do not have it because the internet hampers that connection, then we will want to connect to a drug or behavior that soothes our social needs. Hari (2015b) stated, "The opposite of addiction is not sobriety. The opposite of addiction is connection" (at 13:59 in transcript). Hari's comments are aptly relevant to the addictive nature of the internet, along with video games and social media, all of which provide pseudoconnection while simultaneously socially isolating the user.

REDUCED ATTENTION EQUALS DIVIDED INTENTION

Although there is a large consensus that internet and screen use can be addictive (Aboujaoude et al., 2006; Beard, 2005; Chou et al., 2005; Davis, 2001; Greenfield, 1999a, 1999b, 1999c; Griffiths, 2000a; Griffiths et al., 2016; King & Delfabbro, 2009; Shaw & Black, 2008; Yen et al., 2010; Young, 1998), some simply see the internet as a powerful tool and question it as a diagnosable behavioral or process addiction (Musetti et al., 2016). Young (1996, 2009) and Demir and Kutlu (2018) found that excessive use of the internet for nonacademic and nonprofessional reasons was associated with detrimental effects to academic and occupational performance, respectively. Greenfield (1999a) found that approximately 6% of those who use the internet seem to do so compulsively—often to a point of negative life consequences. Internet addictions are accompanied by a degree of developmental arrest in typical social, occupational, and academic milestones, and often contribute to a delayed launch into independent functioning and living. Patients typically need to relearn real-time living strategies to substitute for the reinforcement that their addiction creates with the typical joys, challenges, and satisfactions balanced living provides.

Still, many questions remain as to appropriate nosology for the labeling of the effects of internet overuse. Although the most popular media terms currently in use are *internet addiction* or *internet use disorder*, other terms used include *pathological internet use, internet-enabled behavior, compulsive internet use, digital media compulsion,* and *virtual addiction* (Greenfield, 1999b). This list is by no means exhaustive but should serve to illustrate the complexity of defining and labeling this clinical phenomenon.

THE DIGITAL DOZEN

In an attempt to illustrate the major features accounting for the addictive aspects of the internet, a list of 12 factors that contribute to the addictive and distractive power of the smartphone and internet was compiled (Greenfield, 2010, 2015):

1. Accessibility: The internet never closes and is readily accessible. This ease of access creates threshold reduction to use and abuse of the internet.

2. Intensity and stimulation (content intoxication): The internet easily accesses stimulating and intoxicating content. The combination of stimulating (addictive) content and the rapid internet delivery mechanism

itself produce synergistic amplification of the accessed content's saliency. The internet is a highly efficient means to impart stimulating content (Greenfield, 2018).

3. Novel and dynamic: The internet is always changing and new. This variability makes the internet novel and psychologically appealing.

4. Dissociation (time distortion): We lose track of time and experience a hyperfocus while distracted from other daily activities, and we cannot judge how much time we are on our screens. It is likely that any substance or behavior that distorts the perception of time may be mood altering (Droit-Volet & Gil, 2009; Greenfield, 1999b). The experience of altered perception of time and space is a ubiquitous experience of the internet medium (Greenfield, 1999a, 2010). The internet content delivery mechanism itself, distinct from the stimulative content found online, alters the perception of time (Greenfield, 2015, 2018).

5. Means of connection: The internet provides a form of social connection. Social media is a good example of *social lite,* in which aspects of social interaction may be experienced but with seemingly less socially nutritive value.

6. Perceived anonymity: The internet gives us a sense of anonymity. Although the internet may actually be the least anonymous form of communication, there is ample perception of the exact opposite (Greenfield, 1999a, 1999b, 1999c).

7. Disinhibition: On the internet, we seem to feel free to be more revealing, and some describe experiencing an accelerated feeling of intimacy (Greenfield, 1999a, 1999b, 1999c). Expressing oneself online seems unencumbered by ego constraints, and the ability to take on an alter ego or persona can itself be intoxicating. Indeed, many people report great pleasure in experiencing and expressing themselves in a completely new way when online.

8. The story without an end: Information is available with no boundaries. The brain tends to try to complete incomplete stories, and the internet always has a new hypertext or link to click, which ultimately provides a time- and attention-draining experience.

9. Cost: The internet provides inexpensive infotainment.

10. Instant gratification: The quicker we get reinforced or rewarded by a substance or a behavior, the more addictive it is. The shorter the latency

between clicking on an internet site, hypertext, or smartphone app and receiving a response, the more addictive it becomes (Allain et al., 2015; Duven et al., 2011).

11. Interactivity: We control and guide the internet search process, and we experience this as being in control; however, it is the internet's reward structure that actually shapes our use and behavior.

12. Variable ratio reinforcement schedule: Dopaminergic innervation fueled by variable operant reinforcements and classically conditioned (notification) responses all have high extinction resistance and addictive potential from anticipatory reward release of dopamine. It is this maybe factor that keeps us in the cycle of endless scrolling and clicking on our screens.

THE SMARTPHONE: A PORTABLE DOPAMINE PUMP

Since my early research, publications, and clinical practice in the late 1990s, the internet has changed dramatically. These changes include broadband and high-speed internet; Wi-Fi; laptops and tablets; digital cell services; faster cell systems; and, of course, the smartphone. The smartphone in 10 short years has become a game changer in terms of ease of access to the internet. It is perhaps the most rapidly adopted technology since the beginning of the industrial revolution; by 2019, 81% of Americans owned a smartphone (Pew Research Center, 2019), and this phenomenon represents the untethering of the internet. The smartphone is quickly becoming the dominant internet access portal in the world ("Percentage of Mobile Device Website Traffic Worldwide From 1st Quarter 2015 to 4th Quarter 2019," n.d.). The accelerated ease of access and short response latency significantly increase the addictive potential of this device (AT&T Inc., 2014; Greenfield, 2017a).

When we hear a ping, buzz, or notification from our smartphones, this activates a reward cue for the possibility of something being there. It is not dissimilar from seeing lights and hearing sounds on a slot machine or video game that let you know that if you play, there could be a reward. This creates a combination of a classical conditioning response loop along with reward conditioning. A neurological connection is created because we associate those notification sounds with the biochemical pleasure response we receive from the elevation of dopamine—and we anticipate the potential of a reward. This unpredictable reward is resistant to extinction (addictive), so we would have to find nothing pleasurable on our smartphone for a long time to naturalistically stop checking our phone (AT&T Inc., 2014; Greenfield, 2017a, 2018).

A smartphone that can be seen or heard elevates cortisol (Heo et al., 2017), and we experience this neurohormonal elevation as discomfort that is easily mediated by checking our phone. When we check our phone and we find something desirable, we likely receive a secondary reinforcing dopamine hit, which further increases the reward loop just as was described in the slot machine analogy (Chun et al., 2018; Greenfield, 2017a). These neurochemical changes contribute to our picking up our smartphone on a consistent basis. The smartphone helps keep us on automatic pilot in a time-distorting, distractible, and vigilant neurophysiologic state, which appears to inhibit us from making intentional choices (Chun et al., 2018; Greenfield, 2017a).

Screen use sequalae can be social isolation, boredom intolerance, and distractibility, which leave us perpetually connected somewhere other than where we are. The internet, and more particularly the smartphone, prevents us from being present centered and is the antithesis of mindful living. When digitally distracted, we become chronically overstimulated and attentionally impaired. Add to this *broadcast capability*, in which our digital culture places diminished value on real-time experiences that are not recorded, posted, shared, and reviewed. Broadcast capability likely creates some level of dopaminergic intoxication that requires continual feeding of the behavior pattern of posting and checking for positive reviews and comments. This seemingly endless cycle is quite purposeful by design from the social media companies to keep our eyes on the screen (Wang, 2017). The result is being removed from visceral, real-time experiences.

Broadcast intoxication (Greenfield, 2016) occurs when an individual experiences aspects of their self-esteem and social valuation by the reviews and reactions of others on social media; this is intoxicating in that it provides clear social reinforcement for our online posting behavior. This phenomenon might be labeled as *reflected self-esteem* (Greenfield, 1986), whereby self-evaluation is based on the echoic reflections from our digital footprint on social media. The social media industry uses social validation looping and social comparison to ensure users will continue to post, read, and review content, which means greater time with eyes on screen. This phenomenon further contributes to the experience of FOMO, or fear of missing out, in which we continually experience and express our lives via social media for fear that we will either be missed or will miss something. Ironically, what we seem to be missing is the experience of our own lives as we become an observer instead of a participant.

WHAT IS DRIVING OUR DISTRACTION?

Compulsive use of our smartphones does not stop when we get into our cars (AT&T, n.d.; AT&T Inc., 2014). People are injured and dying at an alarming rate from using their smartphones while driving (National Highway Traffic Safety Administration, n.d.; National Safety Council, 2012). Perhaps most alarming are recent findings that show it is not only texting that distracts us while we drive but other smartphone functions (AT&T, 2015). Data suggest that 62% of drivers keep their smartphones within reach; 30% who tweet while driving do so "all the time"; 22% of drivers who go to their social networks while behind the wheel report doing so because of addiction; and 27% of people who take videos behind the wheel believe they can film while driving (AT&T, 2015; The Center for Internet and Technology Addiction & AT&T, n.d.).

When we hear the ping of an incoming text, social media update, email, or any pushed notification while driving, our brain receives a small hit of dopamine that impacts our motivation and behavior. The maybe factor—the expectation of a possible reward: Who's texting me? Who tagged me on social media? What is in this email? What did I find in a YouTube video or Google search?—leads to an even higher release of dopamine than the actual content reward itself (AT&T Inc., 2014; Young, 2009).

THE INTERNET AS TOOL OR DESTINATION

Griffiths and colleagues (Griffiths, 2000a; Widyanto & Griffiths, 2006) have argued that people working in internet addiction must differentiate between addiction to the internet versus addiction on the internet. Greenfield (2017b, 2017c, 2018) discussed the phenomenon of *synergistic amplification* in which the simulative value of the content consumed online is separate and distinct from the delivery mechanism; the delivery mechanism is akin to the hypodermic delivering the drug (stimulating content) in our nervous system. The combination of stimulating content delivered via the internet produces an amplified intensity due to stimulatory saliency, immediacy, and variability found online. Furthermore, based on evidence from case examples, Griffiths (2010) contended that excessive internet use may not result in any negative or harmful effects, so those users should not be classified as "addicted." Greenfield (2010, 2018), however, has argued that a significant portion of the definition of *addiction* is based on the deleterious impact of a compulsive pattern of behavior—and that individuals will still engage in such behavior despite the negative consequences. Most "internet addicts," according to Griffiths, may not be addicted to the internet but are using it as a tool to support their access to other addictions. Thus, for example, a person who gambles via the internet may simply have chosen it as the place to carry out their addictive behavior, so that gambling addict is not necessarily an internet addict (Blaszczynski, 2006; King et al., 2011). This argument, though, seems spurious to me because in addiction medicine, we do not separate the hypodermic from the heroin, and the means of ingestion is inexorably connected to the addictive potential of the substance. In the case of the internet, we are looking at the stimulatory potential of the content as the drug and the internet as the delivery mechanism. Furthermore, with all due respect, Griffiths is not a clinician; my theories and techniques for internet addiction treatment have arisen from 23 years of treating this population. From the outset, the issue has been: What do we do with all these patients who are addicted and have significant negative life impacts as a result?

However, some people may conduct behaviors such as cyberstalking only on the internet because it is disinhibiting and offers anonymity (Griffiths, 2000b, 2001; Suler, 2004). Cooper et al. (2000) and Greenfield and Orzack (2002) found that the internet can enable certain sexual behaviors that might not have previously been expressed. In addition, there can be an intensification of previously existing sexual behaviors and paraphilias.

While debate continues about the exact diagnostic criteria that internet addiction or internet use disorder should meet, treatment for internet-related problems in the United States and throughout the world is in great demand (Camardese et al., 2012; Cash et al., 2012; King et al., 2011). In South Korea, China, and Taiwan, estimates are that the prevalence of internet addiction issues among adolescents ranges from 1.6% (K. Kim et al., 2006) to 11.3% (Geng et al., 2006; King et al., 2011); rates in the United States appear to be in similar ranges.

MINDFUL, MODERATED, AND SUSTAINABLE SCREEN USE

Many people overuse their screens and internet technology. Numerous surveys and studies point to upward of 25% of us using our phones and devices to a point of addiction (Common Sense Media, 2016). Hutton et al. (2020) found clear white brain matter changes are directly related to the extent of screen use in preschool-aged children. Kautiainen et al. (2005) found a strong relationship between technology, screen use, and body mass index, and we are most likely to see an increase in sedentary behavior because naturalistic aspects of physical activity have been steadily decreasing over the past 25 years as internet technologies have become more popular. The idea is not to vilify internet and Smartphone use; rather, the goal is to point to the potent time and attention-draining aspects of this indispensable technology.

The internet and all things screen are a powerful psychoactive and addictive medium. I advocate mindful and moderated use of the internet, smartphone, and other digital screen technologies. Hari (2015a) noted that moderated use of addictive substances or behaviors is more easily modulated in individuals who have developed secure attachment and social stimulation or supports. The use of internet and screen technologies for social intimacy and connection is problematic because internet and screen use can be quite isolating and socially disconnecting despite the purported benefits of social connection through social media and screen technologies. Internet and screen use may be the latest drug of choice for many, but addictively numbing and self-medicating are by no means new. The goal should be recognizing the powerful social and neurobiological allure of our screens and making conscious, values-based life choices regarding the time we spend on our screens.

For some, professional help is needed (King et al., 2012; Koo et al., 2011), but for most of us, who may be tech-imbalanced, making digital wellness lifestyle choices may be in order. With national internet use averages increasing, many demographics are reaching extremely high levels of screen time. A Nielsen survey (The Nielsen Company, 2019), for instance, found that Americans 65 and older spent 10 hours a day or more on their screens. For most of us who spend only about 3 hours a day on our combined screen use, which is below the national average, that means about 10 years of our lives are spent on a screen (assuming an 80-year life span). The important question is: What else might we want to do with that time? With the allure and distractive power of the internet, that question perhaps needs to asked sooner rather than later.

REFERENCES

Aboujaoude, E., Koran, L. M., Gamel, N., Large, M. D., & Serpe, R. T. (2006). Potential markers for problematic internet use: A telephone survey of 2,513 adults. *CNS Spectrums, 11*(10), 750–755. https://doi.org/10.1017/S1092852900014875

Adams, P. J. (2016). Switching to a social approach to addiction: Implications for theory and practice. *International Journal of Mental Health and Addiction, 14*(1), 86–94. https://doi.org/10.1007/s11469-015-9588-4

Alexander, B. K. (2012). Addiction: The urgent need for a paradigm shift. *Substance Use & Misuse, 47*(2), 1475–1482. https://doi.org/10.3109/10826084.2012.705681

Allain, F., Minogianis, E. A., Roberts, D. C., & Samaha, A. N. (2015). How fast and how often: The pharmacokinetics of drug use are decisive in addiction. *Neuroscience and Biobehavioral Reviews, 56,* 166–179. https://doi.org/10.1016/j.neubiorev.2015.06.012

American Psychiatric Association. (2013). *Diagnostic and statistical manual of mental disorders* (5th ed.). https://doi.org/10.1176/appi.books.9780890425596

ASAM releases new definition of addiction. (2011). *ASAM News, 26*(3), 1.

AT&T. (n.d.). *It can wait.* http://about.att.com/newsroom/compulsion_research_drivemode_ios_availability.html

AT&T. (2015, May 19). *Smartphone use while driving grows beyond texting to social media, web surfing, selfies, video chatting.* http://about.att.com/story/smartphone_use_while_driving_grows_beyond_texting.html

AT&T Inc. (2014, November 5). *Are you compulsive about texting & driving? Survey says . . . you could be.* https://about.att.com/csr/home/blog/2014/11/are_you_compulsivea.html

Beard, K. W. (2005). Internet addiction: A review of current assessment techniques and potential assessment questions. *Cyberpsychology & Behavior, 8*(1), 7–14. https://doi.org/10.1089/cpb.2005.8.7

Bergmark, K. H., Bergmark, A., & Findahl, O. (2011). Extensive internet involvement—Addiction or emerging lifestyle? *International Journal of Environmental Research and Public Health, 8*(12), 4488–4501. https://doi.org/10.3390/ijerph8124488

Bibbey, A., Phillips, A. C., Ginty, A. T., & Carroll, D. (2015, June). Problematic internet use, excessive alcohol consumption, their comorbidity and cardiovascular and cortisol reactions to acute psychological stress in a student population. *Journal of Behavioral Addictions, 4*(2), 44–52. https://doi.org/10.1556/2006.4.2015.006

Blaszczynski, A. (2006). Internet addiction: In search of an addiction [Editorial]. *International Journal of Mental Health and Addiction, 4,* 7–9. https://doi.org/10.1007/s11469-006-9002-3

Blum, K., Gondré-Lewis, M., Steinberg, B., Elman, I., Baron, D., Modestino, E. J., Badgaiyan, R. D., & Gold, M. S. (2018). Our evolved unique pleasure circuit makes humans different from apes: Reconsideration of data derived from animal studies. *Journal of Systems and Integrative Neuroscience, 4*(1). https://doi.org/10.15761/JSIN.1000191

Bozarth, M. A. (1994). Pleasure systems in the brain. In D. M. Warburton (Ed.), *Pleasure: The politics and the reality* (pp. 5–14). John Wiley & Sons.

Brand, M., Young, K. S., & Laier, C. (2014). Prefrontal control and internet addiction: A theoretical model and review of neuropsychological and neuroimaging findings. *Frontiers in Human Neuroscience, 8,* 375. https://doi.org/10.3389/fnhum.2014.00375

Byun, S., Ruffini, C., Mills, J. E., Douglas, A. C., Niang, M., Stepchenkova, S., Lee, S. K., Loutfi, J., Lee, J. K., Atallah, M., & Blanton, M. (2009). Internet addiction: Metasynthesis of 1996–2006 quantitative research. *Cyberpsychology & Behavior, 12*(2), 203–207. https://doi.org/10.1089/cpb.2008.0102

Camardese, G., De Risio, L., Di Nicola, M., Pizi, G., & Janiri, L. (2012). A role for pharmacotherapy in the treatment of "internet addiction." *Clinical Neuropharmacology, 35*(6), 283–289. https://doi.org/10.1097/WNF.0b013e31827172e5

Cash, H., Rae, C. D., Steel, A. H., & Winkler, A. (2012). Internet addiction: A brief summary of research and practice. *Current Psychiatry Reviews, 8*(4), 292–298. https://doi.org/10.2174/157340012803520513

The Center for Internet and Technology Addiction & AT&T. (n.d.). It Can Wait *compulsion survey*. https://about.att.com/content/dam/snrdocs/It%20Can%20Wait%20Compulsion%20Survey%20Key%20Findings_9%207%2014.pdf

Chou, C., Condron, L., & Belland, J. (2005). A review of the research on internet addiction. *Educational Psychology Review, 17*(4), 363–388. https://doi.org/10.1007/s10648-005-8138-1

Chun, J. W., Choi, J., Cho, H., Choi, M. R., Ahn, K. J., Choi, J. S., & Kim, D. J. (2018). Role of frontostriatal connectivity in adolescents with excessive smartphone use. *Frontiers in Psychiatry, 9*, 437. https://doi.org/10.3389/fpsyt.2018.00437

Comings, D. (2006). S.10.02 Genetics of pathological gambling and substance use disorders. *European Neuropsychopharmacology, 16*(Suppl. 4), S181. https://doi.org/10.1016/S0924-977X(06)70062-1

Common Sense Media. (2016, May 3). *Dealing with devices: The parent–teen dynamic.* https://www.commonsensemedia.org/technology-addiction-concern-controversy-and-finding-balance-infographic

Cooper, A., Boies, S., Maheu, M., & Greenfield, D. (2000). Sexuality and the internet: The next sexual revolution. In F. Muscarella & L. Szuchman (Eds.), *Psychological perspectives on human sexuality: A research based approach* (pp. 519–545). John Wiley and Sons.

Crews, F. T., & Boettiger, C. A. (2009). Impulsivity, frontal lobe and risk for addiction. *Pharmacology Biochemistry and Behavior, 93*(3), 237–247. https://doi.org/10.1016/j.pbb.2009.04.018

Crews, F. T., & Nixon, K. (2009). Mechanisms of neurodegeneration and regeneration in alcoholism. *Alcohol and Alcoholism, 44*(2), 115–127. https://doi.org/10.1093/alcalc/agn079

Davis, R. A. (2001). A cognitive–behavioral model of pathological internet use. *Computers in Human Behavior, 17*(2), 187–195. https://doi.org/10.1016/S0747-5632(00)00041-8

Demir, Y., & Kutlu, M. (2018). The relationship among internet addiction, academic motivation, academic procrastination and school attachment in adolescents. *International Online Journal of Educational Sciences, 10*(5), 315–332. https://doi.org/10.15345/iojes.2018.05.020

de Wit, H. (2009). Impulsivity as a determinant and consequence of drug use: A review of underlying processes. *Addiction Biology, 14*(1), 22–31. https://doi.org/10.1111/j.1369-1600.2008.00129.x

Droit-Volet, S., & Gil, S. (2009). The time–emotion paradox. *Philosophical Transactions of the Royal Society B: Biological Sciences, 364*(1525), 1943–1953. https://doi.org/10.1098/rstb.2009.0013

Duven, E., Müller, K. W., & Wölfling, K. (2011). Internet and computer game addiction—A review of current neuroscientific research. *European Psychiatry, 26*(Suppl. 1), 416. https://doi.org/10.1016/S0924-9338(11)72124-1

Evenden, J. L. (1999). Varieties of impulsivity. *Psychopharmacology, 146*, 348–361. https://doi.org/10.1007/PL00005481

Ferster, C. B., & Skinner, B. F. (1957). *Schedules of reinforcement.* Appleton-Century-Crofts. https://doi.org/10.1037/10627-000

Fu, J., Xu, P., Zhao, L., & Yu, G. (2018). Impaired orienting in youth with internet addiction: Evidence from the Attention Network Task (ANT). *Psychiatry Research, 264*, 54–57. https://doi.org/10.1016/j.psychres.2017.11.071

Geng, Y.-g., Su, L.-y., & Cao, F.-l. (2006). A research on emotion and personality characteristics in junior high school students with internet addiction disorders. *Chinese Journal of Clinical Psychology, 14*, 153–155.

Goudriaan, A. E., Oosterlaan, J., de Beurs, E., & Van den Brink, W. (2004). Pathological gambling: A comprehensive review of biobehavioral findings. *Neuroscience & Biobehavioral Reviews, 28*(2), 123–141. https://doi.org/10.1016/j.neubiorev.2004.03.001

Grant, J. E., Potenza, M. N., Weinstein, A., & Gorelick, D. A. (2010). Introduction to behavioral addictions. *American Journal of Drug and Alcohol Abuse, 36*(5), 233–241. https://doi.org/10.3109/00952990.2010.491884

Greenfield, D. N. (1986). *The effects of self-disclosure, self-esteem, and love/sex attitude similarity on marital satisfaction: A multidimensional analysis* (Publication No. 1986-08) [Doctoral dissertation, Texas Tech University]. Electronic Theses and Dissertations.

Greenfield, D. N. (1999a, August 20–24). *The nature of internet addiction: Psychological factors in compulsive internet use* [Paper presentation]. American Psychological Association 106th Annual Convention, Boston, MA, United States.

Greenfield, D. N. (1999b). Psychological characteristics of compulsive internet use: A preliminary analysis. *Cyberpsychology & Behavior, 2*(5), 403–412. https://doi.org/10.1089/cpb.1999.2.403

Greenfield, D. N. (1999c). *Virtual addiction: Help for netheads, cyberfreaks, and those who love them.* New Harbinger Publications.

Greenfield, D. N. (2010). What makes internet use addictive? In K. Young & C. N. Abreu (Eds.), *Internet addiction: A handbook for evaluation and treatment.* John Wiley and Sons.

Greenfield, D. N. (2015, March 7). *Internet use disorder: Clinical and treatment implications of compulsive internet and video game use in adolescents* [Symposium]. Child & Adolescent Psychiatric Society of Greater Washington Spring Symposium "Addictions and the adolescent brain: Substances, gaming, and the internet," Bethesda, MD, United States.

Greenfield, D. N. (2016, March 17). *Internet and technology addiction: Are we controlling our devices or are they controlling us?* [Keynote address]. The National Association of Social Workers–South Annual Dakota Conference, Sioux Falls, SD, United States.

Greenfield, D. N. (2017a, March 21–23). *Driven to distraction: Why we can't stop using our smartphones when driving* [Keynote address]. 22nd Annual Michigan Traffic Safety Summit, East Lansing, MI, United States.

Greenfield, D. N. (2017b, October 23–28). *Raising Generation D: What parents and clinicians should know about children and compulsive Internet use?* [Panel presentation]. American Academy of Child and Adolescent Psychiatry Annual Meeting, Washington, DC, United States.

Greenfield, D. N. (2017c, October 23–28). *Virtual addiction: Etiological and neurobiological aspects of compulsive internet use* [Panel presentation]. American Academy of Child and Adolescent Psychiatry Annual Meeting, Washington, DC, United States.

Greenfield, D. N. (2018). Treatment considerations in video game addiction and internet use disorder: A qualitative discussion. *Child & Adolescent Psychiatric Clinics, 27*(2), 327–344. https://doi.org/10.1016/j.chc.2017.11.007

Greenfield, D. N., & Orzack, M. H. (2002). The electronic bedroom: Clinical assessment for online sexual problems and internet-enabled sexual behavior. In A. Cooper (Ed.), *Sex and the internet: A guidebook for clinicians* (pp. 129–145). John Wiley & Sons.

Greenfield Recovery Center. (n.d.). *Smartphone addiction*. https://www.greenfieldcenter.com/what-we-treat/smartphone-addiction/

Griffiths, M. D. (2000a). Cyber affairs: A new area for psychological research. *Psychology Review, 7*(1), 28–31.

Griffiths, M. D. (2000b). Excessive internet use: Implications for sexual behavior. *Cyber Psychology & Behavior, 3*(4), 537–552. https://doi.org/10.1089/109493100420151

Griffiths, M. D. (2001). Sex on the internet: Observations and implications for sex addiction. *Journal of Sex Research, 38*(4), 333–342. https://doi.org/10.1080/00224490109552104

Griffiths, M. D. (2010). The role of context in online gaming excess and addiction: Some case study evidence. *International Journal of Mental Health and Addiction, 8*, 119–125. https://doi.org/10.1007/s11469-009-9229-x

Griffiths, M. D., Kuss, D. J., Billieux, J., & Pontes, H. M. (2016). The evolution of internet addiction: A global perspective. *Addictive Behaviors, 53*, 193–195. https://doi.org/10.1016/j.addbeh.2015.11.001

Hari, J. (2015a). *Chasing the scream: The first and last days of the war on drugs*. Bloomsbury.

Hari, J. (2015b, June). *Everything you think you know about addiction is wrong* [Video]. TED Talk. https://www.ted.com/talks/johann_hari_everything_you_think_you_know_about_addiction_is_wrong?language=en&utm_campaign=social&utm_medium=referral&utm_source=facebook.com&utm_content=talk&utm_term=global-social%20issues&fbclid=IwAR3cyTORD-LtNv6dYr2RmtllB-_-HeomFpJVJH8m_CLxh4FPTnex9m9_IhU

Heo, J.-Y., Kim, K., Fava, M., Mischoulon, D., Papakostas, G. I., Kim, M.-J., Kim, D. J., Chang, K.-A. J., Oh, Y., Yu, B.-H., & Jeon, H. J. (2017). Effects of smartphone use with and without blue light at night in healthy adults: A randomized, double-blind, cross-over, placebo-controlled comparison. *Journal of Psychiatric Research, 87*, 61–70. https://doi.org/10.1016/j.jpsychires.2016.12.010

Hutton, J. S., Dudley, J., Horowitz-Kraus, T., DeWitt, T., & Holland, S. K. (2020). Associations between screen-based media use and brain white matter integrity in preschool-aged children. *JAMA Pediatrics, 174*(1), e193869. https://doi.org/10.1001/jamapediatrics.2019.3869

Jorgenson, A. G., Hsiao, R. C.-J., & Yen, C.-F. (2016). Internet addiction and other behavioral addictions. *Child and Adolescent Psychiatric Clinics of North America, 25*(3), 509–520. https://doi.org/10.1016/j.chc.2016.03.004

Jović, J., & Đinđić, N. D. (2011). Influence of dopaminergic system on internet addiction. *Acta Medica Medianae, 50*(1), 60–66. https://doi.org/10.5633/amm.2011.0112

Kautiainen, S., Koivusilta, L., Lintonen, T., Virtanen, S. M., & Rimpelä, A. (2005). Use of information and communication technology and prevalence of overweight and obesity among adolescents. *International Journal of Obesity, 29*, 925–933. https://doi.org/10.1038/sj.ijo.0802994

Kim, K., Ryu, E., Chon, M. Y., Yeun, E.-J., Choi, S.-Y., Seo, J.-S., & Nam, B.-W. (2006). Internet addiction in Korean adolescents and its relation to depression and suicidal ideation: A questionnaire survey. *International Journal of Nursing Studies*, *43*(2), 185–192. https://doi.org/10.1016/j.ijnurstu.2005.02.005

Kim, S. H., Baik, S.-H., Park, C. S., Kim, S. J., Choi, S. W., & Kim, S. E. (2011). Reduced striatal dopamine D2 receptors in people with internet addiction. *NeuroReport*, *22*(8), 407–411. https://doi.org/10.1097/WNR.0b013e328346e16e

King, D. L., & Delfabbro, P. H. (2009). Understanding and assisting excessive players of video games: A community psychology perspective. *Australian Community Psychologist*, *21*(1), 62–74.

King, D. L., Delfabbro, P. H., Griffiths, M. D., & Gradisar, M. (2011). Assessing clinical trials of internet addiction treatment: A systematic review and CONSORT evaluation. *Clinical Psychology Review*, *31*(7), 1110–1116. https://doi.org/10.1016/j.cpr.2011.06.009

King, D. L., Delfabbro, P. H., Griffiths, M. D., & Gradisar, M. (2012). Cognitive-behavioral approaches to outpatient treatment of internet addiction in children and adolescents. *Journal of Clinical Psychology*, *68*(11), 1185–1195. https://doi.org/10.1002/jclp.21918

Knutson, B., Adams, C. M., Fong, G. W., & Hommer, D. (2001). Anticipation of increasing monetary reward selectively recruits nucleus accumbens. *Journal of Neuroscience*, *21*, RC159 (1–5). https://doi.org/10.1523/JNEUROSCI.21-16-j0002.2001

Koo, C., Wati, Y., Lee, C. C., & Oh, H. Y. (2011). Internet-addicted kids and South Korean government efforts: Boot-camp case. *Cyberpsychology, Behavior, and Social Networking*, *14*(6). 391–394. https://doi.org/10.1089/cyber.2009.0331

Koob, G. F., & Volkow, N. D. (2016). Neurobiology of addiction: A neurocircuitry analysis. *The Lancet: Psychiatry*, *3*(8), 760–773. https://doi.org/10.1016/S2215-0366(16)00104-8

Kuss, D. J., & Griffiths, M. D. (2012). Internet and gaming addiction: A systematic literature review of neuroimaging studies. *Brain Sciences*, *2*(3), 347–374. https://doi.org/10.3390/brainsci2030347

Kuss, D. J., & Lopez-Fernandez, O. (2016). Internet addiction and problematic internet use: A systematic review of clinical research. *World Journal of Psychiatry*, *6*(1), 143–176. https://doi.org/10.5498/wjp.v6.i1.143

Lemieux, A., & al'Absi, M. (2016). Chapter 3—Stress psychobiology in the context of addiction medicine: From drugs of abuse to behavioral addictions. *Progress in Brain Research*, *223*, 43–62. https://doi.org/10.1016/bs.pbr.2015.08.001

Liese, B. S., Benau, E. M., Atchley, P., Reed, D., Becirevic, A., & Kaplan, B. (2018). The Self-Perception of Text-Message Dependency Scale (STDS): Psychometric update based on a United States sample. *American Journal of Drug and Alcohol Abuse*, *45*(1), 42–50. https://doi.org/10.1080/00952990.2018.1465572

Moreira, F. A., & Dalley, J. W. (2015). Dopamine receptor partial agonists and addiction. *European Journal of Pharmacology*, *752*, 112–115. https://doi.org/10.1016/j.ejphar.2015.02.025

Musetti, A., Cattivelli, R., Giacobbi, M., Zuglian, P., Ceccarini, M., Capelli, F., Pietrabissa, G., & Castelnuovo, G. (2016). Challenges in internet addiction disorder: Is a diagnosis feasible or not? *Frontiers in Psychology*, *7*, 842. https://doi.org/10.3389/fpsyg.2016.00842

National Highway Traffic Safety Administration. (n.d.). *Distracted driving.* https://www.nhtsa.gov/risky-driving/distracted-driving

National Institute on Drug Abuse. (2018, July). *Drugs, brains, and behavior: The science of addiction.* https://www.drugabuse.gov/publications/drugs-brains-behavior-science-addiction/treatment-recovery

National Safety Council. (2012, April). *Understanding the distracted brain: Why driving while using cell phones is risky behavior* [White paper]. https://www.nsc.org/Portals/0/Documents/DistractedDrivingDocuments/Cognitive-Distraction-White-Paper.pdf?ver=2018-03-09-130423-967

Nesse, R. (2002). Evolution and addiction. *Addiction, 97,* 470–474. https://doi.org/10.1046/j.1360-0443.2002.00086.x

The Nielsen Company. (2019). *The Nielsen total audience report: Q3 2018.* https://www.nielsen.com/wp-content/uploads/sites/3/2019/04/q3-2018-total-audience-report.pdf

Percentage of mobile device website traffic worldwide from 1st quarter 2015 to 4th quarter 2019. (n.d.). Statista. https://www.statista.com/statistics/277125/share-of-website-traffic-coming-from-mobile-devices/

Pew Research Center. (2019, June 12). *Mobile fact sheet.* https://www.pewresearch.org/internet/fact-sheet/mobile/

Ranganath, A., & Jacob, S. N. (2016). Doping the mind: Dopaminergic modulation of prefrontal cortical cognition. *Neuroscientist, 22*(6), 593–603. https://doi.org/10.1177/1073858415602850

Reed, P., Vile, R., Osborne, L. A. Romano, M., & Truzoli, R. (2015). Problematic internet usage and immune function. *PLOS ONE, 10*(8), e0134538. https://doi.org/10.1371/journal.pone.0134538

Ries, R. K., Fiellin, D. A., Miller, S. C., & Saitz, R. (Eds.). (2014). *The ASAM principles of addiction medicine* (5th ed.). Wolters Kluwer.

Robbins, T. W., & Clark, L. (2015). Behavioral addictions. *Current Opinion in Neurobiology, 30,* 66–72. https://doi.org/10.1016/j.conb.2014.09.005

Saah, T. (2005). The evolutionary origins and significance of drug addiction. *Harm Reduction Journal, 2,* Article No. 8. https://doi.org/10.1186/1477-7517-2-8

Schott, B. H., Minuzzi, L., Krebs, R. M., Elmenhorst, D., Lang, M., Winz, O. H., Seidenbecher, C. I., Coenen, H. H., Heinze, H. J., Zilles, K., Düzel, E., & Bauer, A. (2008). Mesolimbic functional magnetic resonance imaging activations during reward anticipation correlate with reward-related ventral striatal dopamine release. *Journal of Neuroscience, 28*(52), 14311–14319. https://doi.org/10.1523/JNEUROSCI.2058-08.2008

Shaw, M., & Black, D. W. (2008). Internet addiction: Definition, assessment, epidemiology and clinical management. *CNS Drugs, 22*(5), 353–365. https://doi.org/10.2165/00023210-200822050-00001

Sohn, S., Rees, P., Wildridge, B., Kalk, N. J., & Carter, B. (2019). Prevalence of problematic smartphone usage and associated mental health outcomes amongst children and young people: A systematic review, meta-analysis and GRADE of the evidence. *BMC Psychiatry, 19,* Article No. 356. https://doi.org/10.1186/s12888-019-2350-x (Correction published 2019, *BMC Psychiatry, 19,* Article No. 397. https://doi.org/10.1186/s12888-019-2393-z)

Suler, J. (2004). The online disinhibition effect. *CyberPsychology & Behavior, 7*(3), 321–326. https://doi.org/10.1089/1094931041291295

Twenge, J. M., Joiner, T. E., Rogers, M. L., & Martin, G. N. (2018). Increases in depressive symptoms, suicide-related outcomes, and suicide rates among U.S.

adolescents after 2010 and links to increased new media screen time. *Clinical Psychological Science, 6*(1), 3–17. https://doi.org/10.1177/2167702617723376

Wang, A. B. (2017, December 12). Former Facebook VP says social media is destroying society with "dopamine-driven feedback loops." *The Washington Post.* https://www.washingtonpost.com/news/the-switch/wp/2017/12/12/former-facebook-vp-says-social-media-is-destroying-society-with-dopamine-driven-feedback-loops/

Widyanto, L., & Griffiths, M. D. (2006). "Internet addiction": A critical review. *International Journal of Mental Health and Addiction, 4,* 31–51. https://doi.org/10.1007/s11469-006-9009-9

Yen, C.-F., Hsiao, R.-C., Ko, C.-H., Yen, J.-Y., Huang, C.-F., Liu, S.-C., & Wang, S.-Y. (2010). The relationships between body mass index and television viewing, internet use and cellular phone use: The moderating effects of socio-demographic characteristics and exercise. *International Journal of Eating Disorders, 43*(6), 565–571. https://doi.org/10.1002/eat.20683

Young, K. (2009). Internet addiction: Diagnosis and treatment considerations. *Journal of Contemporary Psychotherapy, 39,* 241–246. https://doi.org/10.1007/s10879-009-9120-x

Young, K. S. (1996). *Caught in the net: How to recognize the signs of internet addiction—and a winning strategy for recovery.* John Wiley.

Young, K. S. (1998). Internet addiction: The emergence of a new clinical disorder. *CyberPsychology & Behavior, 1*(3), 237–244. https://doi.org/10.1089/cpb.1998.1.237

3 INFORMATION TECHNOLOGY AND ITS IMPACT ON EMOTIONAL WELL-BEING

STEVEN G. GREENING, KACIE MENNIE, AND SEAN LANE

Not too long ago, I (Greening) visited Sachsenhausen while traveling through parts of Western Europe. It was one of the largest German concentration camps during World War II used by the Nazis and later kept open as an internment camp by the Soviets during their occupation of Eastern Germany. Now, Sachsenhausen is a museum that provides a powerful and persistent reminder of a terrible moment in human history. Considering the historical and humane significance of the buildings and artifacts on display, one was left with several somber emotions ranging from sorrow to disgust. My visit occurred on a particularly dreary summer day, which perhaps served to further such negative feelings. Yet despite the "obvious" timbre of Sachsenhausen, numerous tourists posed for gleeful and comedic pictures in front of relics of torture and death. At the time, this struck me as insensitive or ignorant, but there is another possibility. To feel the emotions of Sachsenhausen, one needs to be able to engage in elaborative processing, yet the presence of distracting technology depletes the attentional resources needed for such processing. The presence of a camera and the planning involved in using it were sufficient to impair the ability of some tourists to appraise the experience of Sachsenhausen in a manner that would produce negative emotions.

https://doi.org/10.1037/0000208-004
Human Capacity in the Attention Economy, S. Lane and P. Atchley (Editors)

Modern information technology (IT), like smartphones and tablets, is indeed disrupting cognitive processes like attention (Rosen, Carrier, & Cheever, 2013). But the influence of IT does not stop at how we think; it also affects how we feel. It changes the way we experience and regulate our emotions. IT has this effect not because it impacts differential mechanisms corresponding to cognition and emotion but because cognition and emotion are tremendously intertwined and likely inseparable (Pessoa, 2013). Compounding these issues, IT is purposefully designed to hold our attention because of the incentives provided for both hardware and software designers. For example, more vibrant LED screens grab and maintain our attention even when we should be tired (Cajochen et al., 2011) and even to the point of sleep disruption (Burkhart & Phelps, 2009; Chellappa et al., 2013). Applications (apps), on the other hand, are designed both to capture our attention and, more importantly, to pluck the strings of our emotions. This pulling at our heartstrings includes exploiting our sensitivity for reward and punishment, including socioemotional feedback, in the service of keeping us coming back for more. Many apps are like slot machines designed to reinforce our gambling instinct. These apps use variable reward schedules in which emotional rewards (e.g., photo "likes," new messages, or "leveling up" in games like Clash of Clans or Candy Crush) result only after an unpredictable number of swipes or intervals of time. Still other apps exploit our loss aversion in which we are hostile to losing things like points once we have accrued them. For example, Snapchat encourages kids (and adults) to build up "streaks" of daily interactions with their friends on social media and risk losing all their "streak" points if they go a day without exchanging a message with a given friend (Harris, 2017). Tristan Harris, the former design ethicist for Google, has identified these design incentives, choices, and outcomes in several prescient TED Talks and interviews (Harris, 2014, 2016, 2017) and has called the overall impact of IT on humans a "race to the bottom" in the attention economy. And this race is underpinned by how IT interacts with emotions.

In this chapter, we discuss the impact that IT has on both the experience and the regulation of emotions. We draw on existing evidence from both the behavioral and the neuroscience literature to describe the influence that IT and its capacity for distraction can have on the experience and regulation of emotions. We begin by detailing James Gross's (2014) "modal model" of emotion and subsequently his "process model" of emotion regulation, and briefly discuss some of the neuroscience related to each. Within the frameworks provided by each of Gross's models, we then detail first how IT can impair our emotions and second how IT can potentially improve our emotional well-being. In a final section, we speculate on how inter-actions between IT and socioemotional processes may influence moral decision making.

EMOTIONS AND THEIR NEURAL UNDERPINNINGS

Although a precise definition and an understanding of emotions are still being debated and investigated (Adolphs, 2017; Barrett, 2017), we have, for the purposes of this chapter, adopted Gross's (2014) modal model of emotion. We decided to use Gross's model for three reasons: (a) It includes the sister process model, which considers how emotion regulation can occur by influencing the modal model; (b) much empirical research in psychology, psychophysiology, and cognitive neuroscience has been conduct on the bases of the modal model from which we could draw for the present chapter; and (c) both the modal model and the process model provide frameworks with discrete components that could be discussed in the context of IT. Although we could have chosen a myriad other models or frameworks of emotion, we believe that the arguments, evidence, and suggestions we provide regarding the influence of IT on emotion will remain relevant, even in light of alternative models.

The modal model represents an abstract conceptualization of the emotion generation process beginning with a situation and ending with a response (i.e., a situation–response process; see Figure 3.1). First, there is a situation–person interaction that is potentially significant. The situation may involve a target stimulus along with the surrounding context, and it can be externally (e.g., a bear or a smile) or internally generated (e.g., a negative memory). Second, the situation attracts attention. Third, the meaning of the situation–person interaction is appraised. Last, the product is a complex multimodal response that includes one's physiological, behavioral, and subjective responses. These multimodal responses also serve as feedback to modify the situation (see Figure 3.1). Gross (2014) described the modal model as reflecting person–situation transactions in the generation of emotions, and we examine how IT can influence these transactions by considering each stage in the model. The modal model also forms the basis for the emotion regulation framework discussed in a subsequent section.

We first discuss some of the relevant neuroscience of emotion before we delve into behavioral outcomes predicted by the modal model. In the brain, emotions arise as a function of activated survival circuits, or neural networks, that arbitrate the situation–response process (LeDoux, 2012; Salzman &

FIGURE 3.1. Schematics of Gross's (2014) Modal Model of Emotion Generation and Process Model of Emotion Regulation

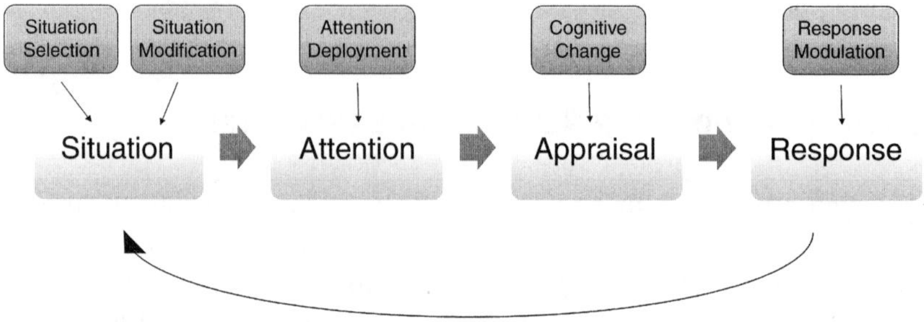

Situation Selection	Situation Modification		Attention Deployment		Cognitive Change		Response Modulation

Situation ▷ Attention ▷ Appraisal ▷ Response

Note. The bottom row in the figure depicts Gross's modal model of emotion, identifying the situation-attention-appraisal-response sequence. The arrow denotes the feedback from a response to a situation. The process model is displayed above the modal model; each emotion regulation process affects a specific node in the modal model. From "Emotion Regulation: Conceptual and Empirical Foundations," by J. J. Gross, in J. J. Gross (Ed.), *The Handbook of Emotion Regulation* (2nd ed., p. 7), 2014, The Guilford Press. Copyright 2014 by The Guilford Press. Adapted with permission.

Fusi, 2010). Consistent with the modal model, the survival circuits that give rise to emotions include regions that represent emotional significance (i.e., valence, intensity, or both), such as the amygdala (see Figure 3.2) but also regions associated with perception, attention, memory, and action. Indeed, the amygdala is richly integrated with the regions associated with these other processes (Bzdok et al., 2013; Greening & Mitchell, 2015; Greening, Lee, & Mather, 2016). The amygdala is a central hub in the emotional circuitry of the mind, including socioemotional stimuli. Functionally, it is involved in the acquisition and expression of learned fear (LaBar et al., 1998; Olsson et al., 2007); is critical in the perception and attention to fearful stimuli like faces or scenes (Kryklywy et al., 2013; Pessoa et al., 2002); and correlates with individual differences in personality characteristics, such as anxiety (Bishop et al., 2007). Although the amygdala is most often associated with fear and negative emotions in general, it is also involved in acquisition and expression of positive emotions. It responds to positive facial expressions (Fusar-Poli et al., 2009), it encodes reward-related associations (Paton et al., 2006), and its activity is positively correlated with individual differences in reward-related processes (Beaver et al., 2006). In addition, direct stimulation of the human amygdala also produces the experience of positive or negative emotion (Lanteaume et al., 2007), further highlighting the amygdala's critical role in both negative and positive emotion generation.

FIGURE 3.2. Brain Regions Involved in Emotional Reactivity and Emotion Regulation

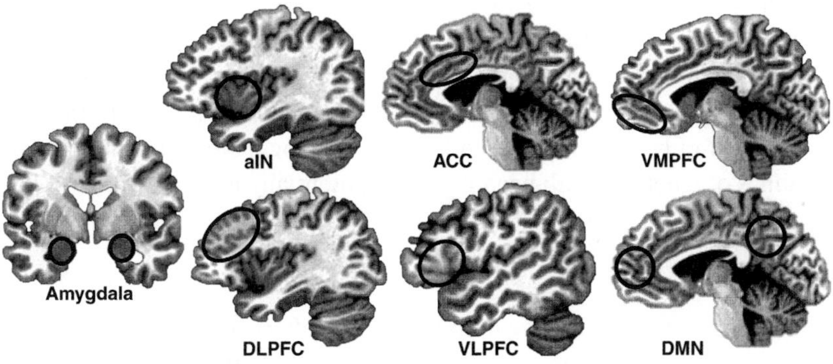

Note. aIN = anterior insula; ACC = anterior cingulate cortex; VMPFC = ventromedial prefrontal cortex; DLPFC = dorsolateral prefrontal cortex; VLPFC = ventrolateral prefrontal cortex; DMN = default mode network.

Although much research has established the importance of the amygdala in emotions, it is not the only brain regions associated with emotion, nor is it always a necessary component to emotion generation. For example, patients with bilateral lesions to the amygdala can experience intense fear evoked by heightened exposure to carbon dioxide (Feinstein et al., 2013), and patients with unilateral amygdala resections report similar daily rates and intensities of subjective positive and negative affect (Anderson & Phelps, 2002). Rather than relying on the amygdala, these examples illustrate the involvement of an extended network of regions referred to as the *core affect circuitry* (Lindquist et al., 2012) that includes the amygdala and other structures, such as the anterior insula (see Figure 3.2), anterior cingulate cortex, and ventromedial prefrontal cortex (see Figure 3.2). Together, they play a domain-general role in the representation of emotions. Thus, to evaluate the impact of IT on emotion, we consider IT's impact not just on the amygdala but on these other areas as well.

EMOTIONS AND IT

Imagine watching your child's school play through the screen on the back of your tablet while you try to record it. You may find that you miss the emotional significance of unexpected scenes of humor—all the while capturing those scenes on video. Or, perhaps you have had the misfortune of trying to have an emotional conversation with a friend only to find that he was not emotionally "moved" by your anecdote because he was busy responding to an email on his smartphone. These examples have been satirized in a YouTube video that has more than 50 million views: *I Forgot My Phone* (deGuzman & Crawford, 2013). We discuss these examples and others in the context of how IT might be influencing the cognitive and neural underpinnings of emotions both negatively and potentially for the better.

Situations

The most fitting place to begin discussing the effect of IT on emotion is to establish that the presence of IT creates situations that negatively affect emotional well-being. Across an increasing number of studies, higher IT use, which includes smartphone and social media engagement, is related to increases in emotional disruptions, such as anxiety and depression (Elhai et al., 2016, 2017; Hoffner & Lee, 2015; Lepp et al., 2014; Panova & Lleras, 2016; Rosen, Whaling, et al., 2013). More recently, Clayton et al. (2015) developed a clever experimental manipulation that points to phones as a cause of anxiety. They found that during "phone separation," participants performed worse on the cognitive task, reported increased unpleasantness, and displayed autonomic markers of increased stress and anxiety, namely, increased heart rate and blood pressure. This feeling associated with being separated from one's phone and therefore potentially missing or being excluded from an important social event has been astutely named FOMO, or fear of missing out (Cheever et al., 2014; Elhai et al., 2016; Rosen, Whaling, et al., 2013). Fear and the associated worry created by potentially threatening situations indeed increase amygdala activity and emotional reactivity, and produce impairments in cognition (Robinson et al., 2012, 2013). The pervasive presence of phones and their notifications in daily life have the potential to create prolonged situations of negative emotions, such as FOMO, that are indistinguishable from veritable situations of threat. Thus, even states such as FOMO can affect one's underlying threat-related neurobiology.

IT can also impact an emotional situation by changing how we experience a situation—as in the earlier example of watching a play through your tablet—to say nothing of the hilarity associated with watching someone trying to use a large tablet as a camcorder given the options available. The aesthetic experience is associated with both positive emotions and corresponding amygdala activity, and there is a link between the aesthetic experience and the perceptual elements in artistic displays (Di Dio et al., 2007). In the present example, rather than the stimuli being real 3D little children running around a stage trying, to the amusement of the crowd, to recall their lines and resist the urge to wave to their parents, the stimuli are 2D figures on a small screen with a limited field of view. Although it has not been empirically confirmed, complementary research has demonstrated that viewing 3D objects can produce significantly different effects both in our brains and behavior compared with pictures of the same objects (Henkel, 2014; Snow et al., 2011, 2014). Viewing our lives through a lens may leave our most precious moments emotionally flat and lacking.

The emotional significance of many situations is often built within a social context. How a given situation makes us feel can depend on who else is present for the situation and how are they behaving. In the brain, areas such as the amygdala that we identified earlier as involved in emotions are also involved in social-emotional responding. The amygdala, for instance, differentially responds more to fearful faces than other expressions and is more active when one's personal space is violated by a person presenting too closely. As an example, one patient, S.M., with bilateral amygdala lesions had impaired social-emotional responding—pronounced deficits recognizing fearful faces such that the patient was unmoved by violations of social norms like "close talking" (Adolphs et al., 1994; Kennedy et al., 2009). Similarly, the anterior insula displays robust activation in social as well as emotional situations (Craig, 2009). In these and other ways, our emotions are closely tied to social interactions.

If IT can create positive social situations, particularly ones we would not have had otherwise, then it can be a force for good as far as emotional well-being is concerned. As globalization continues to increase, many families have become spread across great geographic distances. Apps like Instagram, Facebook, and Skype allow for social situations in which grandparents can see their grandchildren without having to traverse hundreds and thousands of miles. This is one of the observations made by Rosen, Whaling, et al. (2013). In a sample of more than 1,000 participants, they found that more talking on the phone per day and greater numbers of Facebook friends predicted fewer signs of depression and dysthymia. While this is positive news, a cautionary note is needed here because greater time spent online per day and more time spent on Facebook engaged in impression management were predictive of greater signs of depression (Rosen, Whaling, et al., 2013). Engaging in social behavior with one's phone or on social media can improve emotional well-being so long as one avoids various pitfalls. These pitfalls include impression management on social media and other factors meant to hold our attention and keeps us online, such as clickbait headlines, which we describe later.

Attention

Recent functional imaging studies suggest that one role for the amygdala is to signal the presence of an emotionally significant cue in the environment (e.g., the eyes in the identification of fearful facial expressions) and bias attention to that location (Gamer & Büchel, 2009; Han et al., 2012). Although, under some conditions, the amygdala reacts to emotional cues relatively automatically, amygdala reactivity can be attenuated by diverting attention away from emotional cues and toward cognitively demanding tasks (Pessoa et al., 2002). Just like our second scenario that opened the Emotions and IT section, a phone-distracted friend might miss the emotional significance of a social exchange due to a lack of attention. This was borne out in a study by Pessoa et al. (2002) demonstrating that amygdala reactivity to socioemotional cues is suppressed under high levels of distraction compared with when the cues are attended. Such is the effect of distraction in scenarios when a phone-distracted friend responds, "Oh! That's great!" while you are telling them about how horrible your day has been. Although this example demonstrates how one might miss emotional information, there are also instances in which we are overly sensitive to detecting emotional information when doing so is itself a distraction. Bishop et al. (2007) found that individuals with high state anxiety were unable to suppress amygdala activity to low-priority fearful faces when distracting themselves with a low-to-moderate attention task. For example, you may misinterpret every transient facial expression or body posture that your boss makes as negative and as indicative that they disapprove of your work, and all the while, your boss is simply displaying generally neutral expressions, none of which pertains to you. Being sensitive to threat-related signals in the environment may seem beneficial, but such a scenario can become pervasive or even pathological as in the negative biases characteristic of depression or anxiety (Bradley et al., 1995; Greening et al., 2013). Unfortunately, here IT is a potential collaborator in the generation of state anxiety. For example, when we are without our phones, we can experience separation anxiety, or when we are disconnected from social media, we can experience fear of missing out (Cheever et al., 2014). Moreover, the mechanisms of attention control associated with the prefrontal cortex are predictive of individual differences in trait anxiety (Bishop, 2009), suggesting that those who are most anxious may be more susceptible to the attention impairments associated with IT.

The ability to recognize the socioemotional cues of others is second nature for most people, assuming they are not distracted by their smartphones. However, others struggle with emotion processing and related abilities, for example, those with autism spectrum disorder (ASD). Differential patterns of amygdala reactivity and empathy to socioemotional cues, such as faces, are core features of ASD (Swartz et al., 2013). Parts of the amygdala of people with ASD responds significantly less to the eye regions of faces and significantly more to the mouth region (Rutishauser et al., 2013). Such aberrant amygdala reactivity prevents the accurate signaling of emotionally significant cues, leading to disrupted socioemotional perception (Gamer & Büchel, 2009). To overcome such deficits in attention to important socio emotional cues, Gay and Leijdekkers (2014) developed a suite of apps (CaptureMyEmotion) to help children with ASD in the domain of socio-emotional perception, including facial affect recognition. One aspect of the CaptureMyEmotion appears to be its ability to measure how attention is being deployed when looking at facial stimuli. This information could be used to redirect attention to the most diagnostic elements of various facial expressions using the app's interface. Although there is no research into the effectiveness of CaptureMyEmotion, an app such as this could prove useful and is worthy of investigation.

We have already noted that IT devices and their notifications are emotionally relevant, attention grabbing, and potentially harmful. Knowing this, a positive technological innovation would be one that blocks some of the more salient and disruptive aspects of IT, like notifications. Some designers and developers have provided several options for reducing the potential influence of technology on attention. For example, many smartphones and related apps have settings to block disruptive notifications during important times of day, such as during work and bedtime. And applications such as Freedom or Anti-Social for one's computer block distracting websites like Facebook or the internet altogether (see Murray, 2014). If using these inno-vations can stop IT from grabbing our attention and reduce the harm of IT on our emotions, we would consider them good.

Appraisal

As the example that leads this chapter illustrates, being distracted by IT can affect the appraisal process even in what should be the most obvious situations involving relics of torture and stark reminders of human suffering. Even in the most basic of cases of inferring emotional signals from others, the inte-gration of contextual information is important. In several studies, Aviezer and colleagues (2008, 2012) demonstrated that evaluations of emotional faces are influenced by body language. Facial expressions of disgust are readily mistaken as anger if they are paired with angry body language (Aviezer et al., 2008), and facial expressions of loss or frustration are misidentified as expressions of triumph if they are paired with celebratory body language (Aviezer et al., 2012). Brain imaging studies have also demonstrated that the generation of emotional responses generated from top-down verbal descriptions of emotional images involve various aspects of the prefrontal cortex, including the dorsolateral prefrontal cortex (DLPFC; see Figure 3.2). The top-down generation of emotional appraisals also produce stronger emotional reactivity and greater amygdala activity than bottom-up perceptually driven reactivity (McRae et al., 2012; Otto et al., 2014). As powerful as the process of appraisal might be in the generation of affective experiences, the same regions of the prefrontal cortex affecting top-down appraisal are those susceptible to the influence of distraction. Thus, if one is consumed with their technology, the appraisal process can be thwarted even to the extent that it will lead otherwise well-meaning people to smile for photographs with artifacts of torture.

Similar to the potential good that IT can have in the domain of attention in emotion generation, IT can contribute to positive appraisals of emotional situations for those with disorders affecting emotion. The suite of apps called CaptureMyEmotion described previously also has a place in the appraisal niche (Gay & Leijdekkers, 2014). In addition to its potential for helping to improve attention, it also has functions meant to help kids with ASD learn more about emotions and their physiological manifestations. The app is able to provide this physiological information to help kids with ASD become more accurate in the appraisal of their feelings. Similar efforts have been made at developing smartphone apps that combine various types of sensor and implicit user data (e.g., amount of movement in a day) to predict one's mood (LiKamWa et al., 2013). Although it is presently unclear to us how useful such information might be, we speculate it could be leveraged by other user-facing apps. One such app might aim to change how people evaluate their moment-to-moment situations.

Response

Like Alice in a modern-day *Alice in Wonderland*, you have likely been yanked down the rabbit holes of IT and social media, losing tens or hundreds of hours stuck in one social media vortex or echo chamber after another, driven mainly by your outrage and perpetuated by your clicks. Such is the tyranny of the artificial intelligence moderating your use on sites like Facebook whose sole (or soulless) purpose is to keep you on the site as long as possible (Harris, 2017). Much of the clickbait on social media is prioritized by machine learning algorithms or artificial intelligence to optimize user responses. And what do users respond to? Emotion-inducing information. This assuredly includes videos of cats and people falling down, but it also includes videos and articles related to negative emotions like outrage. From the perspective of cognitive and brain sciences, this is not surprising. Emotions influence participant responses during various forms of decision making (Mitchell, 2011), even in rather bizarre yet consequential ways. Judges whose college football team unexpectedly loses dish out harsher sentences in the week following the loss (Eren & Mocan, 2016); similarly, judges are more lenient with their sentences on a full stomach compared with when they are hangry, that is, hungry to the point of anger, before lunchtime (Danziger et al., 2011). In functional imaging studies of reversal learning, responses associated with objects that signal loss or threat are harder to reverse than those associated with positive emotions. While the emotional nature of this task produces activity in parts of the ventromedial prefrontal cortex, it also produces greater activity in parts of the frontal cortex associated with motor planning and execution, such as the inferior frontal gyrus and premotor areas (Greening et al., 2011; Nashiro et al., 2012). It is this relationship between emotional information and how we respond that some apps and social media sites are designed to exploit. The downsides are plentiful and can include greater outrage and anger as well as reduced productivity along with the associated guilt.

IT may have some good to provide in the response phase of the emotion generation process. Panova and Lleras (2016) found that acute phone use in a negative stressful scenario is associated with an initial reduction in negative affect. Thus, the response to pick up one's phone during an emotional situation may provide some benefit. They referred to this as the *security blanket effect.* However, they also observed that longer term use of IT in the emotion generation process was predictive of poor outcomes related to negative emotions and mental health. Like a child who is overly reliant on their security blanket for emotional comfort, at some point, emotional responses and coping must be internalized. This is covered again in the emotion regulation section regarding response modulation.

Feedback

One additional note regarding IT and emotion generation: A feedback loop ties our responses back to the situation, which means that emotional situations can persists, grow, shrink, and change. Although this is relevant to the following section dealing with the regulation of emotions, here, we want to highlight the importance of the response–situation feedback loop. Consider again the scenario described earlier in which a distracted friend misses a joyous anecdote from his friend because he was too busy responding to an email. However, this response frustrates his friend to the point of ire, which morphs what should have been a positive situation into a negative one. Such are the perils of IT in response–situation feedback loops. Consider yourself warned. (Although it seems that abundant possibilities for such a positive-to-negative switch exist, we struggled coming up with examples in which being distracted in social situations will enthuse your friends.)

EMOTION REGULATION AND ITS NEURAL BASIS

Emotion regulation involves the effortful, voluntary, and intentional control of cognition and attention to modulate emotional reactivity. According to Gross's (2014) process model (see Figure 3.1), emotion regulation can affect the emotion generation process at five points along the previously described modal model. The first point occurs before an emotional situation, and the remainder operate at each node of the modal model. Concretely, one can regulate their emotions by *situation selection*, which involves a proactive decision to avoid or indulge a potentially emotional situation; *situation modification*, in which one directly interacts with the physical aspects of a situation to change its impact on emotions; *attentional deployment*, which involves the effortful direction of attention, for example, either externally to nonemotional aspects of a scene or internally to emotionally distinct autobiographical memories; *cognitive change*, which often refers to reappraisal and entails modifying how one thinks of a situation; and *response modulation*, which refers to the modification of the behavioral, physiological, or the subjective feeling of an emotion, or all three, which can include using deep-breathing exercises to slow one's heart rate or masking one's facial expression to suppress an outward display of emotion. Before we—in the next section—unpack how IT can influence each of these processes in turn, we first discuss the relevant neuroscience of emotion regulation.

The cognitive control of emotions emphasized by the process model has been associated with several neural pathways which we describe next. Each of these pathways results in either the direct or indirect modulation of one or several aspects of the core affect regions mentioned earlier, namely, the amygdala, anterior insula, anterior cingulate cortex, or ventromedial prefrontal cortex, or all of them. The planning and effort that one requires to regulate their emotions involve the general processes of attention and cognitive control. This is has been demonstrated in behavioral studies in which working memory and other factors of executive functioning are predictive of emotion regulation ability (Schmeichel & Tang, 2015), and in functional magnetic resonance imaging studies that find that the frontoparietal networks associated with cognitive and attentional control, such as the DLPFC and the ventrolateral prefrontal cortex (see Figure 3.2), are similarly recruited across different regulatory processes, including distraction (e.g., attention deployment to a neutral aspect of a scene), reappraisal, and suppression (Buhle et al., 2014; Goldin et al., 2008; McRae et al., 2010). Despite these similarities, differential neural networks may contribute to the various manifestations of emotion regulation. While there is little research into the neural processes associated with situation selection or modification in the context of emotion regulation, we can speculate some mechanisms based on related work. The construction of imaginary future events, a process likely involved in situation selection and modification, recruits a robust network in medial aspects of the brain, including medial prefrontal cortex and postcentral cortex (Schacter et al., 2012), both of which are part of the default mode network (DMN; see Figure 3.2; Greicius et al., 2003) and are associated with self-relevant processing (Kelley et al., 2002). Attention selection and distraction may work via perceptual mechanisms in early and late sensory areas influenced by bias competition in which goal-relevant stimuli, or stimulus features, are preferentially represented at the expense of goal-irrelevant emotional stimuli (Blair & Mitchell, 2009; Greening, Lee, & Mather, 2014). Conversely, interoceptive attention (e.g., attending to one's heart beat) as used in some mediation strategies involves activation of dorsal insular cortex and posterior cingulate cortex, and reduces activation in sensory areas (Farb et al., 2012). Reappraisal, on the other hand, may rely on semantic mechanisms associated with the temporal and parietal lobes (Buhle et al., 2014) or on value-based mechanisms involving the ventromedial prefrontal cortex (Delgado et al., 2008). A response modulation, or suppression, may rely on mechanisms involved in response inhibition (i.e., reactivity stopping of a prepotent action; Aron, 2011) or reversal learning (i.e., flexibly adapting ones response when response–outcome associations change; e.g., Greening et al., 2011), both of which involve aspects of the ventrolateral prefrontal cortex. In the next section, we cover the role of IT and how it converges with these various forms of emotion regulation and their neural underpinnings.

EMOTION REGULATION AND IT

Consider sitting down to celebrate your birthday. However, you are halfway around the world because you took a job overseas. You are gloomy and alone without your family present for your birthday for the first time in your life. In this moment of despair, you pick up your tablet and video-chat your parents, who instantly cheer you up. On the other hand, envision a scenario in which you are hidden behind the wheel of your car obscured from the view of the person driving the car in front of you. Suddenly the car in front brakes unpredictably, forcing you to screech your car to a halt, narrowly avoiding an accident. Like a child having a tantrum, you then experience an intense and unreasonable anger, or road rage, and begin shouting obscenities out of your window. Although you are normally an even-tempered person who can control your feelings of anger, the mixture of technology (i.e., your car) and emotion has prevented your normally keen faculties of emotion regulation. We consider both these examples and others in the sections that follow as we evaluate the impact that IT can have on our abilities to regulate emotions.

Situation Selection and Modification

As exemplified in the preceding birthday example, humans have a profound ability to select and modify their situations to avoid negative emotions and promote positive ones. However, this ability is easily stifled by distractions. The DMN is a critical region for self-reflection and for imagining future experiences (Schacter et al., 2012). This makes it particularly poised as a cornerstone for situation selection and modification, both of which require the simulation of future self-relevant events to no small degree. Moreover, the DMN is significantly affected in those with depression (Zeng et al., 2012), which indicates that healthy DMN activity is involved in healthy emotion regulation. However, one of the core features of the DMN is that it is suppressed when our attention is diverted to external processes or otherwise distracted (Greicius et al., 2003). Therefore, DMN activity and, therefore, our ability to select and modify our emotional situations are susceptible to disruption by IT. The litany of findings that certain forms of cell phone and social media or internet consumption are associated with increased anxiety, depression, and other forms of emotion dysregulation converge on this conclusion (Clayton et al., 2015; Lepp et al., 2014; Rosen, Whaling, et al., 2013). Moreover, simply being without one's phone or thinking about social media can quickly become a distraction with cognitive and emotional consequences, which one might call withdrawal-like symptoms (Cheever et al., 2014; Clayton et al., 2015; Rosen, Carrier, & Cheever, 2013).

Like the hypothetical birthday boy or girl at the beginning of this section, IT can help you stay in touch with various social networks and provide opportunities for positive situations. One such app that can help to increase social networking and modify one's emotional situations is Meetup. The potential strength of Meetup is that it helps people with shared interests connect for in-person, real-world interactions. This approach may avoid many of the pitfalls associated with high rates of IT consumption (e.g., Lepp et al., 2014; Panova & Lleras, 2016). Some scientists have even developed a "cloud-assisted pillow robot" that allows for social interactions that are simultaneously visual, verbal, and tactile; moreover, they have demonstrated that a traveling mother could use this robot to console her young child while away (Chen et al., 2016).

For those who find in-person interactions difficult, there may also be some benefits to be gained from online communities. In an example that went viral online a few years back, there was a World of Warcraft gamer who died in real life but whose friends were largely other online gamers who had never interacted in person. The friends of the deceased gamer held an online funeral within the World of Warcraft environment to honor their friend and cope with their grief (Luck, 2009).

IT may also be useful in the domain of psychiatric illness. The former director of the National Institute of Mental Health, Tom Insel, recently moved into the private sector with the goal of developing IT that can directly and immediately help patients with mental illness (Dobbs, 2017). His stated goal is to develop a mobile app for predicting and detecting illness onset, and intervening in treatment deployment. In addition, the goal of his app is to integrate with other apps that further aim to provide various services to patients, including those services that enhance social networking. For example, the app PRIME-D has been developed as a depression intervention to modify one's situations using social networking and electronic coaching (Schlosser et al., 2017). These are just two examples of IT with laudable aims to aid in the situation modification processes of emotion regulation. There is, however, a potential downside. Incontrovertible evidence points to a key factor in predicting treatment outcomes of psychotherapy: the humanistic elements (Wampold & Imel, 2015). If patients with psychiatric illness are treated only with IT, then a lack of human contact could undermine the whole enterprise. Video-chatting could provide such a humanistic element, but it could also fall short of a real-life interaction—as might a pillow robot. Nevertheless, given the difficulties associated with accessing a therapist and the cost of psychotherapy, a hybrid approach that combines both real-life and virtual therapy sessions may prove useful.

Attentional Deployment

As you contemplated the road rage scenario, hopefully you developed the sense that IT can become distracting because of its influences on our emotions and because it consumes attentional resources. This is consistent with our earlier arguments. Both emotional and attentional distractions are detrimental to the use of attentional deployment in the services of emotion regulation. First, emotional information has preferential access to perceptual awareness. For example, Stein and Sterzer (2012) found that, compared with neutral faces, happy and angry faces overcome the inhibitory effects of a salient visual distracter more quickly (i.e., shorter suppression duration under continuous flash suppression). Research has also found that goal-irrelevant emotional distracters that robustly activate the amygdala can impair goal-directed behavior (e.g., working memory) by suppressing activity in the frontoparietal attention control regions (Dolcos & McCarthy, 2006). Second, the suppression of distracting emotional stimuli requires greater attentional resources. For example, Amting et al. (2010) found that the suppression of emotional faces from awareness compared with the suppression of neutral faces recruited the frontoparietal attention control regions to a significantly greater extent. Thus, IT can doubly affect attention deployment by impairing frontoparietal attention control systems by way of emotional saliency or directly limiting attentional resources, or most likely both. As we discuss next, IT can be used as a strategy for distracting oneself from emotionally unpleasant situations. This is, however, a potentially double-edge sword. One who finds themselves constantly using IT to distract from negative events may eventually discover they are also distracted from potentially positive events.

Aside from the intrinsically emotional elements of social situations, social interactions can also offer a distraction during tough times. Here, again, apps like Meetup may provide a social distraction. Basic work in cognition and attention finds that social stimuli, like faces, are highly salient and quite distracting for humans (Jenkins et al., 2005; Lavie et al., 2003). Distracting one's self with nonemotional tasks like math problems or other cognitive busywork is also an effective attention deployment strategy. This involves activation of the frontoparietal networks that include the DLPFC and inferior parietal cortex, which can robustly suppress emotional reactivity and related regions, such as the amygdala (Kanske et al., 2011; McRae et al.,

2010; Pessoa et al., 2002). The efficacy of distraction may account for the security blanket effect provided by IT during acute emotional situations (Panova & Lleras, 2016). Therefore, if leveraged in the correct way, IT could provide a healthy distraction from emotionally unpleasant situations.

Cognitive Change

IT can also interact with the cognitive change processes of emotion regulation, such as reappraisal. Hoffner and Lee (2015) found that participants who reported frequently using reappraisal as an emotion regulation strategy were most likely to anticipate missing socioemotional contact and support when they were away from their cell phones. As we have already noted, IT use is both distracting and provokes increases in anxiety. The provocation of anxiety is particularly important given recent research by Raio et al. (2013), which finds that acute stress impairs cognitive reappraisal of emotion regulation. As in the road rage scenario, it might be that the stress of a potentially fatal car accident is acutely debilitating to cognitive change. This debilitation makes it difficult to changes one's thoughts from "the person in the car in front tried to kill me!" to "they stopped suddenly to avoid hitting a pedestrian, and I'm glad we are all safe."

This effect of IT on anxiety and cognitive reappraisal is also problematic of people suffering from mental illness. Growing evidence suggests that the frontoparietal brain regions associated with cognitive reappraisal are either functioning to a lesser extent or have become functionally inefficient in people with depression (Greening, Osuch, et al., 2014; Johnstone et al., 2007). This relationship might also help explain the negative relationship between various types of IT use and depression, anxiety, and other symptoms of mental illness (e.g., Elhai et al., 2017; Panova & Lleras, 2016; Rosen, Whaling, et al., 2013).

One of the challenges associated with cognitive change, or cognitive reappraisal, is that it can sometimes be difficult to generate a reappraisal. When participants are faced with highly arousing emotional stimuli, for example, they prefer avoiding reappraisal in favor of distraction (Sheppes et al., 2011). There are also individual differences in generating reappraisals. Those who are more creative with this process recruit parts of the prefrontal cortex to a greater extent (Fink et al., 2017; Papousek et al., 2017). Based on the relationship among attentional resources, creativity, and emotion regulation, IT may be capable of improving regulation by improving some facet of one's cognitive capacity, namely, emotional working memory. This is precisely what was demonstrated by Schweizer et al. (2013). They had participants train for 20 days using an emotional working memory task and assessed their emotional reappraisal capacity using a pre–post functional magnetic resonance imaging design. Following training, participants in the experimental group (as opposed to a placebo-control group) showed greater reductions in emotional reactivity during the cognitive reappraisal of negative affect along with increased recruitment of the frontoparietal attention control network.

Patients with depression are one group that can find it quite difficult to generate reappraisals. To help patients with depression become better at generating and using reappraisal for emotion regulation, Morris et al. (2015) developed an online platform called Panoply. The goal of the platform is to help people with depression learn and practice emotion reappraisal. It also helps patients generate reappraisal through a crowdsourcing feature that solicits reappraisals from other users. In a promising initial study, Morris and colleagues found that training with Panoply helped significantly reduce symptoms of depression. In this domain of cognitive change, examples of IT like the ones we discussed that work to improve individual capacities for reappraisal could do good.

Response Modulation

In one of the seminal papers on emotion regulation and the process model (Gross, 1998), Gross compared participants who watched disgusting emotional videos while either attempting to reappraise their emotions or suppress them (i.e., by suppressing outward displays of emotion). Although reappraisal reduced sympathetic markers of emotion (e.g., heart rate), suppression increased sympathetic arousal. Moreover, compared with simply watching the videos, reappraisal reduced subjective reports of disgust, and suppression did not. This was also observed in patients with mood and anxiety disorders: Suppression produced greater distress compared with reappraisal (Campbell-Sills et al., 2006). In the brain, whereas reappraisal reduced activity in the amygdala and anterior insula while viewing negative film clips, suppression increased activity in both (Goldin et al., 2008). The sum of this evidence suggests that a singular act of suppression can help regulate one's emotion; however, problems arise when expressive suppression (or bottling up one's emotions) becomes a habitual and favored strategy for emotion regulation (Campbell-Sills et al., 2014). These findings may explain the results of Panova and Lleras (2016), which at first may appear contradictory. They found that smartphone use can initially reduce reactions to stress and negative affect—the security blanket effect. On the other hand, they also discovered that as phone use increased and became excessive, so too was there a positive relationship with mental health problems.

You can probably envisage a time when you were feeling angry or frustrated with a close friend or family member, and rather than berating them to their face, you sat down and wrote a letter or email expressing how you felt. In the end, you may have even decided not to send the letter but were nonetheless left feeling much better having expressed your rage. Response modulation of this sort is a well-documented strategy for emotion regulation that takes many forms, including emotional labeling (Lieberman et al., 2011). *Affective labeling*, or outwardly expressing an emotion, recruits regions of the ventrolateral prefrontal cortex that are also implicated in flexible response modulation during decision making and in cognitive emotion regulation (Lieberman et al., 2007). Thus, a form of IT that encourages emotional labeling could prove quite helpful for people. This is exactly what was demonstrated by Morris and colleagues (Morris & Picard, 2014; Morris et al., 2015). As we described earlier, they developed a web-based platform that aims to help people suffering from depression to better regulate their emotions through reappraisal training. Interestingly, the control condition they used in their initial study had participants engage in an expressive writing online platform. This control condition of expressive writing also produced a significant reduction in depression, although the effect was smaller than for the participants in the reappraisal training group.

Good response modulation can also include using techniques to help regulate physiological responses associated with emotions, such as a racing heart rate and shortness of breath. These include breathing techniques and various forms physical activity and exercise or meditation. Exercise has proven effective at reducing anxiety and improving positive affect (Goldin et al., 2013; Hall et al., 2007; Yeung, 1996). The mechanisms for the effect of exercise on emotion may include the same prefrontal regions associated with attention deployment and cognitive reappraisal (Hall et al., 2007). Insofar as wearable IT such as Fitbit® can increase physical activity in general (Wang et al., 2015), it may also have beneficial side effects for response modulation, emotion regulation, and well-being. Forms of meditation also have beneficial effects for emotion regulation (Goldin et al., 2013). There are myriad apps that provide meditation instruction, for example, the Headspace app. Learning that we can modify our actions to regulate our emotions can be tremendously effective in promoting self-efficacy about emotion regulation (Jazaieri et al., 2014). The opposite scenario of learning that one has no ability to avoid harm (i.e., learned helplessness) can lead to depression like behavior.

MORAL EMOTIONS AND IT

Imagine a train is barreling down the tracks and about to hit five unaware and unreachable rail workers. Now consider two independent scenarios. Although most agree that it is morally permissible to pull a lever, diverting the train to a secondary track, thereby killing a single rail worker to save five, most disagree that it is permissible to push a (sufficiently) large person onto the tracks, thereby again sacrificing one life to save five. Greene and Haidt (2002) distinguished these two scenarios as "impersonal" versus "personal" moral dilemmas, respectively; personal moral dilemmas are highly influenced by our emotional reactions to personally doing harm to another. Such personal dilemmas are associated with greater activity in the DMN compared with impersonal dilemmas (Amit & Greene, 2012; Greene et al., 2001). The personal versus impersonal dichotomy also captures the role for IT in changing our moral intuitions. Technology can dictate the degree to which a decision is personal or impersonal. It is the lever.

Although the train problem seems like an abstract and even unreal philosophical problem, we contend that it has its analog in the real world. Imagine you are the cyberbully writing a malicious message to attract "likes" and contemplating pushing the send button or you are a drone pilot (or their commander) pulling the virtual trigger on a deadly strike from thousands of miles away; in neither case do you see the human face on the receiving end of your decision. For these decisions and many others, it is critical that we understand how IT can influence our moral emotions. On October 7, 2003, 14-year-old Ryan Halligan hanged himself. Ryan's death is perhaps one of the earliest examples of a growing number of teenage suicides related to cyberbullying (LeBlanc, 2012). Unlike traditional in-person bullying, cyberbullying allows bullies to reach their victims any time of day, it provides rapid dissemination of the hateful messaging across large social networks, and it removes having to look your victim in their face and experience the socioemotional impact of the bullying (Ak et al., 2015; Giménez Gualdo et al., 2015). It is this last reason, the disruption of empathy, that is most troubling. Children low on empathy are more likely to perpetuate cyberbullying (Brewer & Kerslake, 2015).

What about the role of IT in the age of drone warfare? In "Confessions of a Drone Warrior," we learn that Brandon Bryant, a drone "sensor," contributed to the deaths of 1,626 enemies in fewer than 6 years (Power, 2013). (The sensor aims the targeting laser for a missile strike but does not fire the drone missile. It is the job of the drone pilot to pull the trigger to fire the missile.) In many of these missions, Bryant was the one targeting and watching on a video screen as human enemies were literally blown to bits. However, Bryant was a cog in the machine of war: a part of the lever being pulled by some distant commander—the orders to fire a missile always coming from higher up the chain of command. The emotional impact of the job on Bryant and his fellow sensors and pilots appears to have contributed to depression, anxiety, alcoholism, and other signs and symptoms of mental illness. Relevant to the current section, while IT allows the commander responsible for ordering the missile strike to make an impersonal decision, Bryant experiences the decision from a personal perspective. This story of Bryant leaves readers with "the creeping sense that screens and cameras have taken some piece of our souls, that we've slipped into a dystopia of disconnection" (Power, 2013, para. 9).

The purpose of this final section is not to make normative claims about what one should or should not do, nor is it to adjudicate the merits of, for example, utilitarian versus deontological moral decision making. Rather we wish to shed light on the fact that insofar as IT influences both emotion and cognition, it will also influence how we decide "right" from "wrong." Although cyberbullying and IT warfare are cautionary tales of the highest order, IT may also be a force for good in moral decision making. As IT permeates the globe, connecting more and more people and allowing greater insights into the various states of the human condition, good may come from the expansion of our moral circle (Singer, 2011) and from movements like effective altruism (MacAskill, 2015).

CONCLUSION

The computer scientist David Krakauer (2016) drew a distinction between complementary cognitive artifacts and competitive cognitive artifacts. Cognitive artifacts are human-made objects that modify our capacity for thought (Norman, 1991). They do so not just by amplifying a certain core ability but by playing on our mind strings of perception, learning, and memory such that they augment and reshape our capabilities. If IT is so distracting that we can tour a concentration camp and holocaust museum without feeling the emotional significance of the place, and if IT so distracts us from the historical significance of such a place that we begin to think, "What's the big deal?," then IT is a competitive cognitive artifact of the worst kind. The aim of this chapter has been to identify both the ways that IT can be a competitive cognitive artifact but also to provide some promise for IT as a complementary cognitive artifact. We can decide to favor and prioritize IT that improves our emotional well-being such that in the absence of it, we retain the benefits. In many cases, the power of IT as a force for good rather than harm is literally held in the palm of our hands.

REFERENCES

Adolphs, R. (2017). How should neuroscience study emotions? By distinguishing emotion states, concepts, and experiences. *Social Cognitive and Affective Neuroscience, 12*(1), 24–31. https://doi.org/10.1093/scan/nsw153

Adolphs, R., Tranel, D., Damasio, H., & Damasio, A. (1994). Impaired recognition of emotion in facial expressions following bilateral damage to the human amygdala. *Nature, 372*(6507), 669–672. https://doi.org/10.1038/372669a0

Ak, Ş., Özdemir, Y., & Kuzucu, Y. (2015). Cybervictimization and cyberbullying: The mediating role of anger, don't anger me! *Computers in Human Behavior, 49*, 437–443. https://doi.org/10.1016/j.chb.2015.03.030

Amit, E., & Greene, J. D. (2012). You see, the ends don't justify the means: Visual imagery and moral judgment. *Psychological Science, 23*(8), 861–868. https://doi.org/10.1177/0956797611434965

Amting, J. M., Greening, S. G., & Mitchell, D. G. (2010). Multiple mechanisms of consciousness: The neural correlates of emotional awareness. *Journal of Neuroscience, 30*(30), 10039–10047. https://doi.org/10.1523/JNEUROSCI.6434-09.2010

Anderson, A. K., & Phelps, E. A. (2002). Is the human amygdala critical for the subjective experience of emotion? Evidence of intact dispositional affect in patients with amygdala lesions. *Journal of Cognitive Neuroscience, 14*(5), 709–720. https://doi.org/10.1162/08989290260138618

Aron, A. R. (2011). From reactive to proactive and selective control: Developing a richer model for stopping inappropriate responses. *Biological Psychiatry, 69*(12), e55–e68. https://doi.org/10.1016/j.biopsych.2010.07.024

Aviezer, H., Hassin, R. R., Ryan, J., Grady, C., Susskind, J., Anderson, A., Moscovitch, M., & Bentin, S. (2008). Angry, disgusted, or afraid? Studies on the malleability of emotion perception. *Psychological Science, 19*(7), 724–732. https://doi.org/10.1111/j.1467-9280.2008.02148.x

Aviezer, H., Trope, Y., & Todorov, A. (2012). Body cues, not facial expressions, discriminate between intense positive and negative emotions. *Science, 338*(6111), 1225–1229. https://doi.org/10.1126/science.1224313

Barrett, L. F. (2017). The theory of constructed emotion: An active inference account of interoception and categorization. *Social Cognitive and Affective Neuroscience, 12*(1), 1–23. https://doi.org/10.1093/scan/nsw154 (Corrigenda published 2017, *Social Cognitive and Affective Neuroscience, 12*(11), p. 1833. https://doi.org/10.1093/scan/nsx060)

Beaver, J. D., Lawrence, A. D., van Ditzhuijzen, J., Davis, M. H., Woods, A., & Calder, A. J. (2006). Individual differences in reward drive predict neural responses to images of food. *Journal of Neuroscience, 26*(19), 5160–5166. https://doi.org/10.1523/JNEUROSCI.0350-06.2006

Bishop, S. J. (2009). Trait anxiety and impoverished prefrontal control of attention. *Nature Neuroscience, 12*(1), 92–98. https://doi.org/10.1038/nn.2242

Bishop, S. J., Jenkins, R., & Lawrence, A. D. (2007). Neural processing of fearful faces: Effects of anxiety are gated by perceptual capacity limitations. *Cerebral Cortex, 17*(7), 1595–1603. https://doi.org/10.1093/cercor/bhl070

Blair, R. J., & Mitchell, D. G. (2009). Psychopathy, attention and emotion. *Psychological Medicine, 39*(4), 543–555. https://doi.org/10.1017/S0033291708003991

Bradley, B. P., Mogg, K., Millar, N., & White, J. (1995). Selective processing of negative information: Effects of clinical anxiety, concurrent depression, and awareness. *Journal of Abnormal Psychology, 104*(3), 532–536. https://doi.org/10.1037/0021-843X.104.3.532

Brewer, G., & Kerslake, J. (2015). Cyberbullying, self-esteem, empathy and loneliness. *Computers in Human Behavior, 48*, 255–260. https://doi.org/10.1016/j.chb.2015.01.073

Buhle, J. T., Silvers, J. A., Wager, T. D., Lopez, R., Onyemekwu, C., Kober, H., Weber, J., & Ochsner, K. N. (2014). Cognitive reappraisal of emotion: A meta-analysis of human neuroimaging studies. *Cerebral Cortex, 24*(11), 2981–2990. https://doi.org/10.1093/cercor/bht154

Burkhart, K., & Phelps, J. R. (2009). Amber lenses to block blue light and improve sleep: A randomized trial. *Chronobiology International, 26*(8), 1602–1612. https://doi.org/10.3109/07420520903523719

Bzdok, D., Laird, A. R., Zilles, K., Fox, P. T., & Eickhoff, S. B. (2013). An investigation of the structural, connectional, and functional subspecialization in the human amygdala. *Human Brain Mapping, 34*(12), 3247–3266. https://doi.org/10.1002/hbm.22138

Cajochen, C., Frey, S., Anders, D., Spati, J., Bues, M., Pross, A., Mager, R., Wirz-Justice, A., & Stefani, O. (2011). Evening exposure to a light-emitting diodes (LED)-backlit computer screen affects circadian physiology and cognitive performance. *Journal of Applied Physiology, 110*(5), 1432–1438. https://doi.org/10.1152/japplphysiol.00165.2011

Campbell-Sills, L., Barlow, D. H., Brown, T. A., & Hofmann, S. G. (2006). Effects of suppression and acceptance on emotional responses of individuals with anxiety and mood disorders. *Behaviour Research and Therapy, 44*(9), 1251–1263. https://doi.org/10.1016/j.brat.2005.10.001

Campbell-Sills, L., Ellard, K. K., & Barlow, D. H. (2014). Emotion regulation in anxiety disorders. In J. J. Gross (Ed.), *The handbook of emotion regulation* (2nd ed., pp. 393–412). The Guilford Press.

Cheever, N. A., Rosen, L. D., Carrier, L. M., & Chavez, A. (2014). Out of sight is not out of mind: The impact of restricting wireless mobile device use on anxiety levels among low, moderate and high users. *Computers in Human Behavior, 37*, 290–297. https://doi.org/10.1016/j.chb.2014.05.002

Chellappa, S. L., Steiner, R., Oelhafen, P., Lang, D., Götz, T., Krebs, J., & Cajochen, C. (2013). Acute exposure to evening blue-enriched light impacts on human sleep. *Journal of Sleep Research, 22*(5), 573–580. https://doi.org/10.1111/jsr.12050

Chen, M., Ma, Y., Hao, Y., Li, Y., Wu, D., Zhang, Y., & Song, E. (2016). Cp-robot: Cloud-assisted pillow robot for emotion sensing and interaction. In J. Wan, I. Humar, & D. Zhang (Eds.), *Industrial IoT technologies and applications: International Conference, Industrial IoT 2016, GuangZhou, China, March 25–26, 2016, Revised Selected Papers* (pp. 81–93). Cham.

Clayton, R. B., Leshner, G., & Almond, A. (2015). The extended iSelf: The impact of iPhone separation on cognition, emotion, and physiology. *Journal of Computer-Mediated Communication, 20*(2), 119–135. https://doi.org/10.1111/jcc4.12109

Craig, A. D. (2009). How do you feel—Now? The anterior insula and human awareness. *Nature Reviews Neuroscience, 10*(1), 59–70. https://doi.org/10.1038/nrn2555

Danziger, S., Levav, J., & Avnaim-Pesso, L. (2011). Extraneous factors in judicial decisions. *Proceedings of the National Academy of Sciences, 108*(17), 6889–6892. https://doi.org/10.1073/pnas.1018033108

deGuzman, C. (Writer), & Crawford, M. (2013, August 22). *I forgot my phone* [Video]. YouTube. https://www.youtube.com/watch?v=OINa46HeWg8

Delgado, M. R., Gillis, M. M., & Phelps, E. A. (2008). Regulating the expectation of reward via cognitive strategies. *Nature Neuroscience, 11*(8), 880–881. https://doi.org/10.1038/nn.2141

Di Dio, C., Macaluso, E., & Rizzolatti, G. (2007). The golden beauty: Brain response to classical and renaissance sculptures. *PLOS One, 2*(11), e1201. https://doi.org/10.1371/journal.pone.0001201

Dobbs, D. (2017, July/August). The smartphone psychiatrist. *The Atlantic, 320*, 78–86. https://www.theatlantic.com/magazine/archive/2017/07/the-smartphone-psychiatrist/528726/

Dolcos, F., & McCarthy, G. (2006). Brain systems mediating cognitive interference by emotional distraction. *Journal of Neuroscience, 26*(7), 2072–2079. https://doi.org/10.1523/JNEUROSCI.5042-05.2006

Elhai, J. D., Dvorak, R. D., Levine, J. C., & Hall, B. J. (2017). Problematic smartphone use: A conceptual overview and systematic review of relations with anxiety and depression psychopathology. *Journal of Affective Disorders, 207*, 251–259. https://doi.org/10.1016/j.jad.2016.08.030

Elhai, J. D., Levine, J. C., Dvorak, R. D., & Hall, B. J. (2016). Fear of missing out, need for touch, anxiety and depression are related to problematic smartphone use. *Computers in Human Behavior, 63*, 509–516. https://doi.org/10.1016/j.chb.2016.05.079

Eren, O., & Mocan, N. (2016). *Emotional judges and unlucky juveniles. American Economic Journal: Applied Economics, 10*(3), 171–205. https://doi.org/10.1257/app.20160390

Farb, N. A. S., Segal, Z. V., & Anderson, A. K. (2012). Attentional modulation of primary interoceptive and exteroceptive cortices. *Cerebral Cortex, 23*(1), 114–126. https://doi.org/10.1093/cercor/bhr385

Feinstein, J. S., Buzza, C., Hurlemann, R., Follmer, R. L., Dahdaleh, N. S., Coryell, W. H., Welsh, M. J., Tranel, D., & Wemmie, J. A. (2013). Fear and panic in humans with bilateral amygdala damage. *Nature Neuroscience, 16*(3), 270–272. https://doi.org/10.1038/nn.3323

Fink, A., Weiss, E. M., Schwarzl, U., Weber, H., de Assunção, V. L., Rominger, C., Schulter, G., Lackner, H. K., & Papousek, I. (2017). Creative ways to wellbeing: Reappraisal inventiveness in the context of anger-evoking situations. *Cognitive, Affective & Behavioral Neuroscience, 17*(1), 94–105. https://doi.org/10.3758/s13415-016-0465-9

Fusar-Poli, P., Placentino, A., Carletti, F., Landi, P., Allen, P., Surguladze, S., Benedetti, F., Abbamonte, M., Gasparotti, R., Barale, F., Perez, J., McGuire, P., & Politi, P. (2009). Functional atlas of emotional faces processing: A voxel-based meta-analysis of 105 functional magnetic resonance imaging studies. *Journal of Psychiatry & Neuroscience, 34*(6), 418–432.

Gamer, M., & Büchel, C. (2009). Amygdala activation predicts gaze toward fearful eyes. *Journal of Neuroscience, 29*(28), 9123–9126. https://doi.org/10.1523/JNEUROSCI.1883-09.2009

Gay, V., & Leijdekkers, P. (2014). Design of emotion-aware mobile apps for autistic children. *Health and Technology, 4*(1), 21–26. https://doi.org/10.1007/s12553-013-0066-3

Giménez Gualdo, A. M., Hunter, S. C., Durkin, K., Arnaiz, P., & Maquilón, J. J. (2015). The emotional impact of cyberbullying: Differences in perceptions and experiences as a function of role. *Computers & Education, 82*, 228–235. https://doi.org/10.1016/j.compedu.2014.11.013

Goldin, P., Ziv, M., Jazaieri, H., Hahn, K., & Gross, J. J. (2013). MBSR vs aerobic exercise in social anxiety: fMRI of emotion regulation of negative self-beliefs. *Social Cognitive and Affective Neuroscience, 8*(1), 65–72. https://doi.org/10.1093/scan/nss054

Goldin, P. R., McRae, K., Ramel, W., & Gross, J. J. (2008). The neural bases of emotion regulation: Reappraisal and suppression of negative emotion. *Biological Psychiatry, 63*(6), 577–586. https://doi.org/10.1016/j.biopsych.2007.05.031

Greene, J., & Haidt, J. (2002). How (and where) does moral judgment work? *Trends in Cognitive Sciences, 6*(12), 517–523. https://doi.org/10.1016/S1364-6613(02)02011-9

Greene, J. D., Sommerville, R. B., Nystrom, L. E., Darley, J. M., & Cohen, J. D. (2001). An fMRI investigation of emotional engagement in moral judgment. *Science, 293*(5537), 2105–2108. https://doi.org/10.1126/science.1062872

Greening, S. G., Finger, E. C., & Mitchell, D. G. (2011). Parsing decision making processes in prefrontal cortex: Response inhibition, overcoming learned avoidance, and reversal learning. *NeuroImage, 54*(2), 1432–1441. https://doi.org/10.1016/j.neuroimage.2010.09.017

Greening, S. G., Lee, T. H., & Mather, M. (2014). A dual process for the cognitive control of emotional significance: Implications for emotion regulation and disorders of emotion. *Frontiers in Human Neuroscience, 8*, 253. https://doi.org/10.3389/fnhum.2014.00253

Greening, S. G., Lee, T. H., & Mather, M. (2016). Individual differences in anticipatory somatosensory cortex activity for shock is positively related with trait anxiety and multisensory integration. *Brain Sciences, 6*(1), 2. Advance online publication. https://doi.org/10.3390/brainsci6010002

Greening, S. G., & Mitchell, D. G. (2015). A network of amygdala connections predict individual differences in trait anxiety. *Human Brain Mapping, 36*(12), 4819–4830. https://doi.org/10.1002/hbm.22952

Greening, S. G., Osuch, E. A., Williamson, P. C., & Mitchell, D. G. (2013). Emotion-related brain activity to conflicting socio-emotional cues in unmedicated depression. *Journal of Affective Disorders, 150*(3), 1136–1141. https://doi.org/10.1016/j.jad.2013.05.053

Greening, S. G., Osuch, E. A., Williamson, P. C., & Mitchell, D. G. (2014). The neural correlates of regulating positive and negative emotions in medication-free major depression. *Social Cognitive and Affective Neuroscience, 9*(5), 628–637. https://doi.org/10.1093/scan/nst027

Greicius, M. D., Krasnow, B., Reiss, A. L., & Menon, V. (2003). Functional connectivity in the resting brain: A network analysis of the default mode hypothesis. *Proceedings of the National Academy of Sciences, 100*(1), 253–258. https://doi.org/10.1073/pnas.0135058100

Gross, J. J. (1998). Antecedent- and response-focused emotion regulation: Divergent consequences for experience, expression, and physiology. *Journal of Personality and Social Psychology, 74*(1), 224–237. https://doi.org/10.1037/0022-3514.74.1.224

Gross, J. J. (2014). Emotion regulation: Conceptual and empirical foundations. In J. J. Gross (Ed.), *The handbook of emotion regulation* (2nd ed., pp. 3–20). The Guilford Press.

Hall, E. E., Ekkekakis, P., & Petruzzello, S. J. (2007). Regional brain activity and strenuous exercise: Predicting affective responses using EEG asymmetry. *Biological Psychology, 75*(2), 194–200. https://doi.org/10.1016/j.biopsycho.2007.03.002

Han, T., Alders, G. L., Greening, S. G., Neufeld, R. W. J., & Mitchell, D. G. V. (2012). Do fearful eyes activate empathy-related brain regions in individuals with callous traits? *Social Cognitive and Affective Neuroscience, 7*(8), 958–968. https://doi.org/10.1093/scan/nsr068

Harris, T. (2014, December 16). *Distracted? Let's make technology that helps us spend our time well* [Video]. TEDx Talks, YouTube. https://www.youtube.com/watch?v=jT5rRh9AZf4

Harris, T. (2016, November 15). *Anti-blind: How do you see what you can't see?* [Video]. TEDx Talks, YouTube. https://www.youtube.com/watch?v=TQPD2Qv4Pls

Harris, T. (2017, May 22). *How tech uses unethical tricks to addict us (Pt. 1)* [Video]. The Rubin Report. https://www.youtube.com/watch?v=qsUrOmwI82I&t=912s

Henkel, L. A. (2014). Point-and-shoot memories: The influence of taking photos on memory for a museum tour. *Psychological Science, 25*(2), 396–402. https://doi.org/10.1177/0956797613504438

Hoffner, C. A., & Lee, S. (2015). Mobile phone use, emotion regulation, and well-being. *Cyberpsychology, Behavior, and Social Networking, 18*(7), 411–416. https://doi.org/10.1089/cyber.2014.0487

Jazaieri, H., McGonigal, K., Jinpa, T., Doty, J. R., Gross, J. J., & Goldin, P. R. (2014). A randomized controlled trial of compassion cultivation training: Effects on mindfulness, affect, and emotion regulation. *Motivation and Emotion, 38*(1), 23–35. https://doi.org/10.1007/s11031-013-9368-z

Jenkins, R., Lavie, N., & Driver, J. (2005). Recognition memory for distractor faces depends on attentional load at exposure. *Psychonomic Bulletin & Review, 12*(2), 314–320. https://doi.org/10.3758/BF03196378

Johnstone, T., van Reekum, C. M., Urry, H. L., Kalin, N. H., & Davidson, R. J. (2007). Failure to regulate: Counterproductive recruitment of top-down prefrontal-subcortical circuitry in major depression. *Journal of Neuroscience, 27*(33), 8877–8884. https://doi.org/10.1523/JNEUROSCI.2063-07.2007

Kanske, P., Heissler, J., Schönfelder, S., Bongers, A., & Wessa, M. (2011). How to regulate emotion? Neural networks for reappraisal and distraction. *Cerebral Cortex, 21*(6), 1379–1388. https://doi.org/10.1093/cercor/bhq216

Kelley, W. M., Macrae, C. N., Wyland, C. L., Caglar, S., Inati, S., & Heatherton, T. F. (2002). Finding the self? An event-related fMRI study. *Journal of Cognitive Neuroscience, 14*(5), 785–794. https://doi.org/10.1162/08989290260138672

Kennedy, D. P., Gläscher, J., Tyszka, J. M., & Adolphs, R. (2009). Personal space regulation by the human amygdala. *Nature Neuroscience, 12*(10), 1226–1227. https://doi.org/10.1038/nn.2381

Krakauer, D. (2016, September 6). *Will A.I. harm us? Better to ask how we'll reckon with our hybrid nature.* Nautilus. http://nautil.us/blog/will-ai-harm-us-better-to-ask-how-well-reckon-with-our-hybrid-nature

Kryklywy, J. H., Nantes, S. G., & Mitchell, D. G. (2013). The amygdala encodes level of perceived fear but not emotional ambiguity in visual scenes. *Behavioural Brain Research, 252,* 396–404. https://doi.org/10.1016/j.bbr.2013.06.010

LaBar, K. S., Gatenby, J. C., Gore, J. C., LeDoux, J. E., & Phelps, E. A. (1998). Human amygdala activation during conditioned fear acquisition and extinction: A mixed-trial fMRI study. *Neuron, 20*(5), 937–945. https://doi.org/10.1016/S0896-6273(00)80475-4

Lanteaume, L., Khalfa, S., Régis, J., Marquis, P., Chauvel, P., & Bartolomei, F. (2007). Emotion induction after direct intracerebral stimulations of human amygdala. *Cerebral Cortex, 17*(6), 1307–1313. https://doi.org/10.1093/cercor/bhl041

Lavie, N., Ro, T., & Russell, C. (2003). The role of perceptual load in processing distractor faces. *Psychological Science, 14*(5), 510–515. https://doi.org/10.1111/1467-9280.03453

LeBlanc, J. C. (2012, October 20–23). *Cyberbullying and suicide: A retrospective analysis of 22 cases* [Paper presentation]. American Academy of Pediatrics AAP Experience National Conference & Exhibition, New Orleans, LA, United States. https://aap.confex.com/aap/2012/webprogrampress/Paper18782.html

LeDoux, J. (2012). Rethinking the emotional brain. *Neuron, 73*(4), 653–676. https://doi.org/10.1016/j.neuron.2012.02.004

Lepp, A., Barkley, J. E., & Karpinski, A. C. (2014). The relationship between cell phone use, academic performance, anxiety, and satisfaction with life in college students. *Computers in Human Behavior, 31,* 343–350. https://doi.org/10.1016/j.chb.2013.10.049

Lieberman, M. D., Eisenberger, N. I., Crockett, M. J., Tom, S. M., Pfeifer, J. H., & Way, B. M. (2007). Putting feelings into words. *Psychological Science, 18*(5), 421–428. https://doi.org/10.1111/j.1467-9280.2007.01916.x

Lieberman, M. D., Inagaki, T. K., Tabibnia, G., & Crockett, M. J. (2011). Subjective responses to emotional stimuli during labeling, reappraisal, and distraction. *Emotion, 11*(3), 468–480. https://doi.org/10.1037/a0023503

LiKamWa, R., Liu, Y., Lane, N. D., & Zhong, L. (2013, June). Moodscope: Building a mood sensor from smartphone usage patterns. In H.-H. Chu, P. Huang, R. R. Choudhury, & F. Zhao (Program Chairs), *MobiSys '13: Proceeding of the 11th Annual International Conference on Mobile Systems, Applications, and Services* (pp. 389–402). Association for Computing Machinery. https://doi.org/10.1145/2462456.2464449

Lindquist, K. A., Wager, T. D., Kober, H., Bliss-Moreau, E., & Barrett, L. F. (2012). The brain basis of emotion: A meta-analytic review. *Behavioral and Brain Sciences, 35*(3), 121–143. https://doi.org/10.1017/S0140525X11000446

Luck, M. (2009). Crashing a virtual funeral: Morality in MMORPGs. *Journal of Information, Communication and Ethics in Society, 7*(4), 280–285. https://doi.org/10.1108/14779960911004516

MacAskill, W. (2015). *Doing good better: Effective altruism and a radical new way to make a difference.* Guardian Faber.

McRae, K., Hughes, B., Chopra, S., Gabrieli, J. D., Gross, J. J., & Ochsner, K. N. (2010). The neural bases of distraction and reappraisal. *Journal of Cognitive Neuroscience, 22*(2), 248–262. https://doi.org/10.1162/jocn.2009.21243

McRae, K., Misra, S., Prasad, A. K., Pereira, S. C., & Gross, J. J. (2012). Bottom-up and top-down emotion generation: Implications for emotion regulation. *Social Cognitive and Affective Neuroscience, 7*(3), 253–262. https://doi.org/10.1093/scan/nsq103

Mitchell, D. G. (2011). The nexus between decision making and emotion regulation: A review of convergent neurocognitive substrates. *Behavioural Brain Research, 217*(1), 215–231. https://doi.org/10.1016/j.bbr.2010.10.030

Morris, R. R., & Picard, R. (2014). Crowd-powered positive psychological interventions. *Journal of Positive Psychology, 9*(6), 509–516. https://doi.org/10.1080/17439760.2014.913671

Morris, R. R., Schueller, S. M., & Picard, R. W. (2015). Efficacy of a web-based, crowd-sourced peer-to-peer cognitive reappraisal platform for depression: Randomized controlled trial. *Journal of Medical Internet Research, 17*(3), e72. https://doi.org/10.2196/jmir.4167

Murray, S. (2014, December 17). The internet restriction apps that help improve productivity. *The Guardian.* https://www.theguardian.com/small-business-network/2014/dec/17/internet-restriction-apps-productivity

Nashiro, K., Sakaki, M., Nga, L., & Mather, M. (2012). Differential brain activity during emotional versus nonemotional reversal learning. *Journal of Cognitive Neuroscience, 24*(8), 1794–1805. https://doi.org/10.1162/jocn_a_00245

Norman, D. (1991). Cognitive artifacts. In J. M. Carroll (Ed.), *Designing interaction: Psychology at the human–computer interface* (pp. 17–38). Cambridge University Press.

Olsson, A., Nearing, K. I., & Phelps, E. A. (2007). Learning fears by observing others: The neural systems of social fear transmission. *Social Cognitive and Affective Neuroscience, 2*(1), 3–11. https://doi.org/10.1093/scan/nsm005

Otto, B., Misra, S., Prasad, A., & McRae, K. (2014). Functional overlap of top-down emotion regulation and generation: An fMRI study identifying common neural substrates between cognitive reappraisal and cognitively generated emotions. *Cognitive, Affective & Behavioral Neuroscience, 14*(3), 923–938. https://doi.org/10.3758/s13415-013-0240-0

Panova, T., & Lleras, A. (2016). Avoidance or boredom: Negative mental health outcomes associated with use of information and communication technologies depend on users' motivations. *Computers in Human Behavior, 58,* 249–258. https://doi.org/10.1016/j.chb.2015.12.062

Papousek, I., Weiss, E. M., Perchtold, C. M., Weber, H., de Assunção, V. L., Schulter, G., Lackner, H. K., & Fink, A. (2017). The capacity for generating cognitive reappraisals is reflected in asymmetric activation of frontal brain regions. *Brain Imaging and Behavior, 11*(2), 577–590. https://doi.org/10.1007/s11682-016-9537-2

Paton, J. J., Belova, M. A., Morrison, S. E., & Salzman, C. D. (2006). The primate amygdala represents the positive and negative value of visual stimuli during learning. *Nature, 439*(7078), 865–870. https://doi.org/10.1038/nature04490

Pessoa, L. (2013). *The cognitive–emotional brain: From interactions to integration.* The MIT Press. https://doi.org/10.7551/mitpress/9780262019569.001.0001

Pessoa, L., McKenna, M., Gutierrez, E., & Ungerleider, L. G. (2002). Neural processing of emotional faces requires attention. *Proceedings of the National Academy of Sciences, 99*(17), 11458–11463. https://doi.org/10.1073/pnas.172403899

Power, M. (2013, October 23). Confessions of a drone warrior. *GQ.* https://www.gq.com/story/drone-uav-pilot-assassination

Raio, C. M., Orederu, T. A., Palazzolo, L., Shurick, A. A., & Phelps, E. A. (2013). Cognitive emotion regulation fails the stress test. *Proceedings of the National Academy of Sciences, 110*(37), 15139–15144. https://doi.org/10.1073/pnas.1305706110

Robinson, O. J., Charney, D. R., Overstreet, C., Vytal, K., & Grillon, C. (2012). The adaptive threat bias in anxiety: Amygdala-dorsomedial prefrontal cortex coupling and aversive amplification. *NeuroImage, 60*(1), 523–529. https://doi.org/10.1016/j.neuroimage.2011.11.096

Robinson, O. J., Krimsky, M., & Grillon, C. (2013). The impact of induced anxiety on response inhibition. *Frontiers in Human Neuroscience, 7*, 69. https://doi.org/10.3389/fnhum.2013.00069

Rosen, L. D., Carrier, L. M., & Cheever, N. A. (2013). Facebook and texting made me do it: Media-induced task-switching while studying. *Computers in Human Behavior, 29*(3), 948–958. https://doi.org/10.1016/j.chb.2012.12.001

Rosen, L. D., Whaling, K., Rab, S., Carrier, L. M., & Cheever, N. A. (2013). Is Facebook creating "iDisorders"? The link between clinical symptoms of psychiatric disorders and technology use, attitudes and anxiety. *Computers in Human Behavior, 29*(3), 1243–1254. https://doi.org/10.1016/j.chb.2012.11.012

Rutishauser, U., Tudusciuc, O., Wang, S., Mamelak, A. N., Ross, I. B., & Adolphs, R. (2013). Single-neuron correlates of atypical face processing in autism. *Neuron, 80*(4), 887–899. https://doi.org/10.1016/j.neuron.2013.08.029

Salzman, C. D., & Fusi, S. (2010). Emotion, cognition, and mental state representation in amygdala and prefrontal cortex. *Annual Review of Neuroscience, 33*, 173–202. https://doi.org/10.1146/annurev.neuro.051508.135256

Schacter, D. L., Addis, D. R., Hassabis, D., Martin, V. C., Spreng, R. N., & Szpunar, K. K. (2012). The future of memory: Remembering, imagining, and the brain. *Neuron, 76*(4), 677–694. https://doi.org/10.1016/j.neuron.2012.11.001

Schlosser, D. A., Campellone, T. R., Truong, B., Anguera, J. A., Vergani, S., Vinogradov, S., & Arean, P. (2017). The feasibility, acceptability, and outcomes of PRIME-D: A novel mobile intervention treatment for depression. *Depression and Anxiety, 34*(6), 546–554. https://doi.org/10.1002/da.22624

Schmeichel, B. J., & Tang, D. (2015). Individual differences in executive functioning and their relationship to emotional processes and responses. *Current Directions in Psychological Science, 24*(2), 93–98. https://doi.org/10.1177/0963721414555178

Schweizer, S., Grahn, J., Hampshire, A., Mobbs, D., & Dalgleish, T. (2013). Training the emotional brain: Improving affective control through emotional working memory training. *Journal of Neuroscience, 33*(12), 5301–5311. https://doi.org/10.1523/JNEUROSCI.2593-12.2013

Sheppes, G., Scheibe, S., Suri, G., & Gross, J. J. (2011). Emotion-regulation choice. *Psychological Science, 22*(11), 1391–1396. https://doi.org/10.1177/0956797611418350

Singer, P. (2011). *The expanding circle: Ethics, evolution, and moral progress.* Princeton University Press. https://doi.org/10.1515/9781400838431

Snow, J. C., Pettypiece, C. E., McAdam, T. D., McLean, A. D., Stroman, P. W., Goodale, M. A., & Culham, J. C. (2011). Bringing the real world into the fMRI scanner: Repetition effects for pictures versus real objects. *Scientific Reports, 1*, Article No. 130. https://doi.org/10.1038/srep00130

Snow, J. C., Skiba, R. M., Coleman, T. L., & Berryhill, M. E. (2014). Real-world objects are more memorable than photographs of objects. *Frontiers in Human Neuroscience, 8*, 837. https://doi.org/10.3389/fnhum.2014.00837

Stein, T., & Sterzer, P. (2012). Not just another face in the crowd: Detecting emotional schematic faces during continuous flash suppression. *Emotion, 12*(5), 988–996. https://doi.org/10.1037/a0026944

Swartz, J. R., Wiggins, J. L., Carrasco, M., Lord, C., & Monk, C. S. (2013). Amygdala habituation and prefrontal functional connectivity in youth with autism spectrum disorders. *Journal of the American Academy of Child & Adolescent Psychiatry, 52*(1), 84–93. https://doi.org/10.1016/j.jaac.2012.10.012

Wampold, B. E., & Imel, Z. E. (2015). *The great psychotherapy debate: The evidence for what makes psychotherapy work.* Routledge. https://doi.org/10.4324/9780203582015

Wang, J. B., Cadmus-Bertram, L. A., Natarajan, L., White, M. M., Madanat, H., Nichols, J. F., Ayala, G. X., & Pierce, J. P. (2015). Wearable sensor/device (Fitbit One) and SMS text-messaging prompts to increase physical activity in overweight and obese adults: A randomized controlled trial. *Telemedicine Journal and e-Health, 21*(10), 782–792. https://doi.org/10.1089/tmj.2014.0176

Yeung, R. R. (1996). The acute effects of exercise on mood state. *Journal of Psychosomatic Research, 40*(2), 123–141. https://doi.org/10.1016/0022-3999(95)00554-4

Zeng, L.-L., Shen, H., Liu, L., Wang, L., Li, B., Fang, P., Zhou, Z., Li, Y., & Hu, D. (2012). Identifying major depression using whole-brain functional connectivity: A multivariate pattern analysis. *Brain, 135*(5), 1498–1507. https://doi.org/10.1093/brain/aws059

PART **II** HOW INFORMATION
TECHNOLOGY
INFLUENCES
COGNITION AND
PERFORMANCE

4 INFORMATION TECHNOLOGY AND LEARNING

KEVIN YEE

As discussed in earlier chapters, the disruptive effects of technology on attention could be seen as ultimately positive or negative (or, more realistically, as sophisticated combinations of both), with much depending on the disposition and agenda of the viewer. Yet taken as a whole, the disruptions caused by technology do not form a recognizable pattern or directionality. The disruption comes without any particular teleological outcome or preordained conclusion. As a consequence, observers of technology have always had an easy time convincing the public that the stamp they wish to imprint on any given new technology is both plausible and inevitable, a move that has applied equally well to tech adherents and doubters. Adherents are then unceremoniously dismissed as too-ardent believers, while doubters are derisively labeled Luddites (Frischmann, 2019). Because both sides have a seemingly realistic claim to credibility, both receive equal attention in the popular media. And thus any given new technology could manage to inhabit two quantum states of existence simultaneously—apparently both the best and the worst thing to happen to humankind, if one merely interprets various receptions to that technology juxtaposed to each other.

https://doi.org/10.1037/0000208-005
Human Capacity in the Attention Economy, S. Lane and P. Atchley (Editors)

Nowhere is this dichotomy more apparent than in teaching and learning, and in particular in higher education. In this chapter, I examine in closer detail the nature and depth of this dichotomy, beginning with setting the context of technology as a disruptive force. Then, I investigate the direct benefits to learning that technology can bring, before shifting to more subtle indirect benefits to learning. Then, I examine the potential detriments to learning in the higher education environment that technology might enable, and conclude by exploring potential next steps and implementation of technology policy, given the double-edged sword of benefits and detriments posed by many technology tools.

I will begin by examining the potential disruption that one particular technology promised to bring to education. With that disruption came confusion and uncertainty as the potential benefits and detriments were weighed. While breathless accounts about the next imminent paradigm shift have always abounded in every discipline, I will focus on one prominent disruptor in higher education. College professors were seemingly about to become extinct, if one believed the breathless articles about the potential of massive open online courses (MOOCs) when they burst onto the higher-education landscape like a wrecking ball in the early 2010s (Christensen et al., 2010). And indeed MOOCs were seen as the ultimate disruptor, a term applied to multiple industries whenever the true potential of the internet laid bare the inadequacies of the old business model while simultaneously dangling newer, more lucrative replacement ones if we would just reach out for them. The classic example invoked is the bookstore model, chained by history and slow-moving executives to a bricks-and-mortar distribution model, and thus one ripe for disruption by the internet (Sultan & Jamal Al-Lail, 2015). The online model, quickly dominated by Amazon.com, leveraged supply-chain, distribution, and fulfillment inefficiencies to replace them with more-efficient models driven by internet access and cloud capabilities, and therefore could undercut pricing of retail stores so completely that most of them went out of business. Thus, "disruption by the internet" came to mean the decimation of an older industry by a newer one that leveraged the advantages of the online environment, and MOOCs were seen as the flaming chariot coming for the stolid higher education industry (Sultan & Jamal Al-Lail, 2015). Surely, the existence of free online lectures and entire courses would mean that students would rebel against paying for products so similar to what was available without cost only a few clicks away?

As it turned out, MOOCs did not provide the radical disruption once expected (or feared, based on one's role within the academy). Ironically, it was the nature of attention itself that ultimately proved to be MOOCs' Achilles heel. At colleges and universities, paid tuition is required, but this results in a credential in the form of course credit, and eventually a diploma, could be earned. Higher education has always faced the criticism that it provides, at the end of the day, a transaction: To oversimplify, cash is paid and a diploma results—or at least these are the most visible inputs and outputs of the system. MOOCs were either free, meaning there was no transaction at all, or when they cost money they provided in return a standalone credential of a single course that almost never led to a diploma. The result was the removal of any significant reward for completing the course, and without that kind of external motivation, MOOC completion rates of around 6.5% on average (Handoko et al., 2019) paled in comparison with those for similar online courses taken by students in degree-seeking programs. Without a strong connection to a larger credential, MOOCs could not provide enough intrinsic value to sustain the work required for completion, and they are no longer seen as any kind of threat to enrollment at colleges and universities (Sultan & Jamal Al-Lail, 2015).

By no means are MOOCs unique in briefly wearing the crown of ultimate EdTech disruptor only to lose it soon after. Another high-profile example might be seen in Second Life, which occurred half a decade earlier. Second Life was a "persistent" online universe, like a video game that kept running even if all the human users logged off in that corner of the digital world. Human users showed up here as avatars—cartoon versions of themselves they could customize and thus have a "second life" different from their own real one. Tech enthusiasts were enamored by the concept. Unlike a video game, however, there were so few rules and central tasks that the environment actually seemed barren, and users were confused about what to do. In the end, only true enthusiasts embraced it, and it remained a niche market (Au, 2013), despite several universities paying top dollar for a persistent presence in Second Life, usually in the form of an electronic campus and the shells of several digital buildings that had few practical uses.

In fact, the arrival of new technologies on the college campus is such a common occurrence that the New Media Consortium's Horizon Project (https://www.nmc.org/nmc-horizon/) provides annual updates of which websites, apps, programs, and hardware are likely to make waves in the education space in the short, medium, and long-term time frame. These Horizon reports are often mapped onto the famous Gartner Hype Cycle diagram (https://www.gartner.com/en/research/methodologies/gartner-hype-cycle). As its name implies, this diagram in its generic form charts technology hype and adoption across time into five periods: the technology trigger, the peak of inflated expectations, the trough of disillusionment, the slope of enlightenment, and the plateau of productivity. Virtually all technologies that eventually became successful had passed through stages such as these, with overinflated expectations naturally leading toward user disillusionment. Not every technology makes it to the next stage, but tech adopters find it useful to see which of the resilient products make it to the later stages of productive implementation. The danger, of course, is that the dizzying array of contenders dissuades all but the truest believers to wait until winners emerge, and the demise of early favorites could lead to long-term disillusionment even among so-called early adopters. When that happens, the pace of technological adoption slows in general, with fewer early adopters providing the yeoman's work of advancing promising ideas to later stages (Chen & Han, 2019).

DIRECT BENEFITS TO LEARNING

With the stage thus set for a realistic sense of tech survivability, we turn our attention to the direct benefits to learning that can occur when technology is used to capture attention. The example most likely to leap immediately to mind, given their visibility and adoption over the past dozen years, are student response systems, often informally referred to as "clickers." Originally only available as dedicated hardware, clickers are wireless devices synced to a central receiving station at the teacher's console. About the size of a television remote control, clickers allow students to "vote" on questions posed on screen, with real-time results available to juice discussion about the distribution of responses. The system is also frequently used for accountability, such as attendance-taking or taking short auto-grading quizzes via multiple choice. Since the early 2010s, vendors have increasingly prioritized an app-and-subscription model wherein the students "bring your own device" (BYOD) to the classroom and use existing WiFi or cellular connections to join the class electronically (Imazeki, 2014). While laptops and tablets clearly qualify, students prefer smartphones by a wide majority that is growing annually (Stowell, 2015).

Clickers, through both dedicated hardware and BYOD options, have been demonstrated to bring great benefit to student learning (Pertanika, 2018; Rana et al., 2016). They provide a mechanism to interrupt long lectures and thus reset attention, they offer formative feedback to both students and instructors about the status of learning, and they can be used before particular lecture points to awaken interest and drive novelty. Most important, clickers represent a way to include more assessments in the course, leading to increased retrieval practice and thus have a positive effect on long-term memory (Hubbard & Couch, 2018).

Even in the face of such positive gains, however, lurk dangers and down-sides. The mere presence of physical clickers does not guarantee increases in learning. It matters quite a lot how the clickers are used pedagogically (Rana et al., 2016). My direct experience with faculty over a decade working in a teaching center is that clickers may not bring benefits if used only sparingly in a long lecture, or used with multiple-choice questions perceived by the audience as boring or too easy. I have tracked very negative student reaction to instructors who require students to buy clickers simply for the purpose of taking attendance, generating no pedagogical lift at all. Such instances are rare, however, and the vast majority of clicker implementations are well conceived and offer solid learning gains for students.

The use of smartphones for BYOD clicker applications offers a boost to learning in ways beyond engagement and formative feedback. Since so many students own smartphones, instructors may feel more confident than they otherwise would in requiring the use of BYOD clickers for graded assessments, such as quizzes (or even micro-quizzes as short as single questions). Whether through the use of clicker software, requiring a software license that costs students money, or the use of the institution's learning management system (LMS) directly, which does not cost extra money, instructors have access to self-grading quizzes that do not add substantively to their workload. The ease and convenience of BYOD quizzing encourages the use of more frequent assessment, which has been shown conclusively to result in large gains in student learning (Kyriazi, 2015). These are particularly pronounced when frequent quizzing is combined with a cumulative approach, forcing students to study constantly in a manner consistent with spaced retrieval practice and interleaving, which refers to mixing together practice of different kinds of content rather than separating them out. The combination of spaced retrieval practice and interleaving supports even stronger long-term recall (Dunlosky et al., 2013).

The BYOD mentality is not restricted to clickers and clicker-like activity such as LMS quizzing, and students have long made use of consumer electronics to aid in their college career. Particularly in prior decades this was visible in the use of handheld recording devices such as mini-cassette recorders, deployed in a lecture hall to enable students to review a lecture later. While re-listening to entire lectures has dubious learning potential when compared with the much stronger results gained by reviewing notes (Willingham, 2009), captured lectures can aid in cases where students need disability accommodations. In addition, these lectures can provide a fruitful way to revisit particularly problematic concepts (so long as the attempt is targeted, rather than a misguided attempt to study by listening to everything again). Smartphones can perform similar recording functions by app, but this practice is less commonly employed these days, possibly to avoid draining the smartphone battery. It is also relevant to note that some classrooms and lecture halls are now equipped with dedicated hardware and software solutions that enable lecture capture controlled by the instructor, and made available online, obviating the need for students to capture their own recordings in such instances (Lambert et al., 2019).

Lecture capture as controlled by the instructor typically combines both audio and video feeds, and in fact often includes a second video feed of the instructor's screen. Students watching the recordings later—they are not typically broadcast in real time—can listen to the audio while they watch the instructor or the instructor's screen, or switch between them at will. The video component provides a clear advantage over audio-only recordings of prior decades mentioned previously, as many explanations of advanced concepts include visual components to aid understanding. While all students may derive a benefit from a repeated viewing of difficult concepts, as noted, there are certain audiences that benefit even more, such as students needing accommodations for disabilities and those who struggle with the language of instruction. The latter group often spends class time mentally translating the instructor's words, doubling the time taken to process new information and much more frequently bumping up against the limits of cognitive load, resulting in note-taking of lesser quality and overall diminished comprehension of the lecture, especially in more subtle nuances (Watt et al., 2014). The increased time-on-task offered by lecture capture, particularly in combination with the ability to pause or even slow down the playback, provides these students with a critical learning aid (Watt et al., 2014).

Even beyond clicker-like apps and LMS quizzing, smartphones offer access to knowledge that can be profitably leveraged in several different ways during classroom time. To offer a simple example, a student hearing a fact during lecture may use a smartphone to look up additional background, context, and details about the fact, which increases the likelihood of consolidation of the knowledge to long-term memory. This is due to the simple fact that items with additional context and relevance are encoded to long-term memory at a higher rate than those without such context and relevance (Willingham, 2009). While such enhanced web searches would be possible postlecture if earmarked by pen-and-paper note-taking, they are less likely to occur since the opportunity and momentary motivation are lost. In this sense, the ubiquity of smartphones provides a key benefit, as does the trend toward instant gratification.

Instead of relying only on student initiative, faculty may opt to leverage smartphones and their access to the internet directly into their pedagogical choices. Examples might include a modern version of "webquest," a technique popularized in K–12 education. In a webquest, students are given a problem and told to use the internet to find answers themselves. In this fashion, the students are given first-order access to information in lieu of lecture (Alibec & Sandiuc, 2015). Since the students access the information directly, the experience is more direct and thus of higher impact. Clearly, there is an engagement boost to student-driven web searching over instructor-provided knowledge; attention issues typical of lectures would be largely mitigated (Sahlström et al., 2019). Instructors might realize the largest gains when smartphone searches are done as part of an intentional group activity rather than isolated practice (Sahlström et al., 2019).

Among the universe of smartphone apps are those geared to assist learning, and many incorporate learning science into their design. DuoLingo, for example, elevates the typical flashcard experience when learning a foreign language to a new level by placing an emphasis on interleaving and distributed practice. Rather than rely on the user to return to old material, it is placed in front of them in irregularly spaced intervals to require retrieval and encourage consolidation into long-term memory.

Nor do the benefits of technology for learning end with student-held technology. Computer labs have long held promise for student learning, with applications as wide-ranging as digital homework problems to audio-lingual practice for foreign language. One particularly promising pedagogy for classroom computers is the emporium model, in which students complete homework (usually math) directly in the computer lab during assigned classroom time (Sears et al., 2019). Lectures occur in the same time slot, sometimes in a different lecture hall on designated lecture days, but when the computer practice time is the priority, students work on problems individually but have access to the instructor or teaching assistants when questions arise. The method's effectiveness rests on access to experts during the application period. In a traditional learning environment, students complete the homework privately, away from the classroom environment and likely sequestered as individuals. In the emporium model, they need only signal for help whenever they are stuck. The attentional benefits to an emporium classroom are obvious; with students engaged in personal effort in every minute, coupled with an incentive to finish early and thus leave early, they face few temptations to attempt task-switching. The resulting time-on-task is high quality from an attentional point of view (Sears et al., 2019).

The emporium model can be used in conjunction with the concept of a flipped class, but not all flipped classrooms use an emporium approach during face-to-face time. In general, a flipped class can be thought of as one that moves lectures and content delivery completely away from classroom time, relying on recorded lectures or assigned texts to deliver content and dedicating class time wholly to practice activities such as worked examples, group work, case studies, and scenario analysis. The technology component is usually visible in the asynchronous capture of lectures. These may take the form of screencasts (often simple audio overlaid atop PowerPoint slides) or more elaborate lecture captures from a non-classroom environment, such as the instructor's office.

Flipped classrooms offer enhanced learning possibilities primarily through the shift in usage of classroom time (Freeman et al., 2014). Since content delivery has been moved to homework and online delivery, classroom time is freed up to focus on applications, group activities, and other means of direct practice. The resulting classroom time is highly engaging for students, with an emphasis on interactivity and active (rather than passive) engagement with the material. Active learning methods have been long shown to be superior to student learning than more passive modes, such as lecture (Freeman et al., 2014).

Emerging technologies offer the promise of learning gains, primarily through efforts to make the learning personal and customized. Virtual reality (VR) and augmented reality (AR) use visors or specialty glasses to view the environment differently. In VR, visors block the normal world entirely, and the wearer sees a digitally created environment directly in front of their eyes, with the gaze shifting with every turn of the head, as if the wearer were actually inside the computer-generated reality. In AR, glasses are worn that allow the everyday world to be seen, but computer-generated images on the glasses augment the reality. AR can also be accomplished through a screen, such as using the camera function of a smartphone to show the real world, plus a digitally added item, as in the popular game Pokemon Go. They are highly engaging because they offer immediacy and interactivity, and the experience as a whole is perceived as novel because these sensory perceptions are not part of normal human existence. This novelty drives interest, attention, and thus engagement with the material (Chin et al., 2018), though questions remain about whether the learning gains will be as pronounced once the novelty wears off (Merchant et al., 2014). Most technologies offered boosts in attention when they were novel, but not every technology persists as a learning aid. Both VR and AR efforts are presently too expensive to produce to become mainstream, and as a result are primarily visible for targeted audiences, such as advanced medical training (Merchant et al., 2014).

The Holy Grail of learning has, for years, been the promise of personalization (Grant & Basye, 2014). Educators dream of making an array of learning tools available to students with the goal that each person would navigate the learning ecosystem differently and in tune with their skills and preferences. One person might opt for a video-heavy content delivery, for instance, while another would prefer the more sedate and reflective route of reading similar material. While personalized learning environments could include low-tech solutions, the preferred investigation has been toward adaptive learning—software solutions that adjust the depth and difficulty of the material presented/quizzed based on the user's ongoing score to quiz questions. For example, a student faring poorly on division problems will see more division problems on the ongoing quizzes until this improves, while one scoring highly on division will progress more rapidly to advanced topics.

In this manner, adaptive testing alters one of the fundamental tenets of higher education over the past century (Hoffman & Omsted, 2018). Our current system relies on the "Carnegie hour," a concept popularized by educational theorist Dale Carnegie. In his view, college subjects should be taught over a 15-week semester with three hours of class meetings per week. Essentially, students are given a set amount of material to learn, and required to engage with the material for a minimum number of hours to constitute a semester. But of course students are not homogeneous automatons, and in reality some of them are ahead of the average, and others behind, as the semester begins. What results by holding the content and the time commitment constant is that the variable is the amount of learning—those needing little get little, and those needing a lot get a lot. Adaptive learning offers a technique for a different way of thinking about learning often called competency-based learning: as soon as a student has demonstrated competency, they should advance to the next topics, regardless of how long or short a time this requires (Hoffman & Omsted, 2018). For competency-based learning, the content and the amount of learning are held constant, resulting in time required becoming the variable (Hoffman & Omsted, 2018). Because of the nature of adaptive testing, where students are constantly presented with exactly the right amount of challenge, there are fewer opportunities for boredom or gaps in attention.

Competency-based education has many advantages and interest in it is growing, but it does not mesh easily with the systems built over decades that require completion measured in established blocks of time, rather than demonstrated mastery of a discipline (Hoffman & Omsted, 2018). Adaptive testing and other technologies that support competency-based education may yet prove to be the ultimate disruptive technology in education, if they succeed in changing the rules of what higher-education degrees mean and how to earn them.

INDIRECT BENEFITS TO LEARNING

The pervasiveness of technology in our everyday lives makes it inevitable that there are some applications that could be looked at to provide indirect benefits to learning in higher education. While not necessarily tools that instructors would deploy directly within the class environment, some provide ancillary support, whereas others simply supply inspiration because the level of participant engagement is so high that one seeks to emulate it, or at least identify and adapt the principles that generate such strong engagement.

Consider the case of informal learning, in which something other than education is the primary goal—usually it's purely entertainment and diversion—but some manner of learning might occur during the playing of the game as a by-product. One example might be modern middle-schoolers learning the names of medieval weapons such as halberds, poleaxes, and glaives while playing video games. The early exposure via video games creates both awareness and latent curiosity, which can be capitalized upon in college coursework, for example, or when the student visits an armory in Europe. The resulting learning in the armory is made much stronger by the earlier groundwork laid by the video games (Scolari & Contreras-Espinosa, 2019).

Games turn out to be excellent models of activities that hold attention amid high engagement—the ideal fertile ground for learning (Deterding et al., 2011). While games have been part of human existence for centuries, the more recent popularity of video games (a category that here includes phone-based game apps) has led to increased attention from academia (Bowman, 2018), in particular as applied to self-determination theory of psychological predictors of behavioral outcomes (Deci & Ryan, 1985; Deci & Ryan, 2002). Clearly, higher education has much to learn from video games about sustained attention.

Early attempts to classify and categorize games led to large taxonomies such as Yu-Kai Chou's Octalysis Framework, in which domains such as meaning, empowerment, social influence, unpredictability, avoidance, scarcity, ownership, and accomplishment can be mapped onto any game, with dozens of subfactors such as virtual goods, beginner's luck, and friending. But the principles of video game engagement—those things that make games fun—could be narrowed to just a few suggestions: display progress, maximize competition, calibrate difficulty properly, provide diversions from the main goal, and employ narratives when possible (Bell, 2017).

The first two principles, displaying progress and maximizing competition, could take the form of leaderboards within a class, but this needs to be done carefully as privacy laws dictate student grades not be displayed (Bell, 2017). One suggested workaround is to use student groups as the unit of competition, such as seen in the house competitions of the Harry Potter books and movies (Bell, 2017). Setting the difficulty of the game means that its tasks must offer just the right level of challenge: too easy and it will become boring, too difficult and it becomes frustrating and might lead to capitulation. Even professional game designers struggle to ensure that the difficulty of a game is properly balanced. In the educational environment, instructors designing games must not only attempt to gauge a game for the imagined students, they must also play-test the game with students at an equivalent level. The next principle, diversions, might be thought of as "side games" or "side quests" that are not related to or required for the main competition, but are fun in their use of novelty. One example might be to use mouseover-text on embedded photographs of the course LMS pages, offering funny or quirky commentary simply to encourage the engagement. Finally, many successful games occur in a situated context of an overarching narrative, which aligns with a hunger for stories that appears to be universal (Bell, 2017). To "gamify" a college course is to employ these principles in concert with each other in a semester-long competition, perhaps by posting a leaderboard (or badges) in the LMS, and having groups compete to win new badges toward eventual final victory.

Gamification has made its way into society beyond education. Hybrid cars often display the miles-per-gallon (MPG) equivalent at the current driving speed, thus subtly encouraging slower overall speeds to maximize the MPG running total (Canali, 2016). This kind of competition-against-self still tracks progress and leverages our competitive nature, and in the final analysis provides the impetus to alter behavior. Nudge theory (Sunstein et al., 2019) has long pointed to the ability to encourage certain behaviors through the use of reminders and setting of certain defaults. In the college classroom, faculty can nudge students to take action through the use of announcements or other targeted tools within the LMS, but there are useful third-party tools as well. The smartphone app Remind offers a way to nudge students to study at times across the workweek, thus distributing retrieval practice and encouraging long-term memory formation. Popular study apps like Quizlet offer digital versions of flashcards that combine interleaving and spaced practice as well, while using nudges to gamify the overall experience and make studying seem more fun.

The final element of gamification mentioned previously, employing narrative where possible, builds on ideas long known about the connection between memory and story, first noticed with oral traditions of epic heroes but applicable to narratives as small as urban legends (Heath & Heath, 2007).

Indeed, many video game players can list historical details about fictional kingdoms from these games, since they find the stories so engrossing and the facts simply lodge in their memory as a by-product. While there are certainly lessons to be drawn from this experience for all teaching, some technology applications are tying narratives directly to the learning outcomes. For example, the company Embodied Labs is using VR capabilities to bring alive the lived experiences of patients with certain medical conditions such as Alzheimer's disease or macular degeneration (Burge, 2018). Because the story is the crucial component of the VR experience, students see through the eyes of the patient and gain an understanding of the empathies required as a medical professional. As with any narrative, this VR experience enables enhanced memory of the details (Gabana et al., 2017), made all the more deep by the near-reality nature of the VR.

POTENTIAL DETRIMENTS TO LEARNING

While the principles of gamification can be intentionally applied to virtually any teaching context to make the material more engaging, it should be noted that the effect is almost exclusively realized as extrinsic motivation; in effect, "tricking" students to learn rather than engaging their interest in the material toward stirring intrinsic motivation. Long term, students with intrinsic motivations learn more deeply than those who are stimulated only externally (Deci, 1975), so gamification must be viewed with this caveat.

Each of the technologies mentioned previously has similar caveats as well. For instance, faculty are often reluctant to investigate and adopt new technologies because they have learned, often through trial and error, that the impermanence of new applications creates duplication or even wasted effort (Wang & Wang, 2009). Instructors seeking to pursue a technology to interrupt PowerPoint videos with questions, to require interactivity and attention, saw only a few options recently, the best of which was a plug-in called Office Mix. Once Office Mix shut down in 2018, the interactive presentations were useless, and instructors had to start over with a different application to achieve similar results (Alexander, 2018). Over more than a dozen years in faculty development, I have seen that the uncertainty about the shelf life of any technology makes some faculty hesitant to incorporate any unproven technology, which means the benefits listed previously are never realized for those students.

Occasionally, the nature of the technology itself creates unintended consequences that can be negative for learning. Lecture capture offers advantages for students to rewatch the live performance on video, which can certainly aid with comprehension of difficult subjects but also enables students to skip the lecture the first time and watch the recording instead (Owston et al., 2011). These should not be mistaken for identical learning environments. Without the presence of the instructor, students are more likely to task-switch and be engaged with distractions (Lepp et al., 2019), which has a measurably negative effect on their learning (Judd & Kennedy, 2011). Some lecture capture software platforms offer the ability to watch the recording at different speeds, including double-speed. Students may watch the lecture at double-speed to advance more quickly through verbal fillers, but an increase in input speed does not result in increased processing speeds, so that the net result is that students are able to get less out of a video watched at double speed (Varnavsky, 2016). Finally, the well-known phenomenon of the illusion of mastery (Brown et al., 2014) is not ameliorated by lecture capture. The illusion of mastery occurs when students have few questions during lecture because the expert explanation suffices for passive understanding, but struggle later to actively produce the same conceptual understanding once away from the expert. Even a second viewing of a difficult subject provides only passive understanding, creating the risk that the illusion of mastery has been reproduced rather than true mastery.

Even clickers, in the form of the BYOD variations, have a dark side. The vast majority of students use their personal smartphone as their clicker device. Since the phones are now physically in their hands, those phones become temptations far greater than when they were safely tucked away in pockets and handbags (Kuznekoff & Titsworth, 2013). After the required voting is accomplished, the usual downtime while classmates finish their voting is easily filled by opening a different app. This leads to the ironic situation that the attempt to engage them and focus their attention on the content actually causes misdirected attention and distraction for some students.

That many of today's students are addicted to social media seems little in doubt (O'Donnell & Epstein, 2019). These modern day "weapons of mass distraction" (Rosen, 2012) affect attention negatively within the lecture hall in several overlapping ways. The first is the impact on the task-switching student directly, whose attention necessarily shifts away from the class to focus on the device. The second is the impact on students located near (and especially behind) the task-switching student. Because our eyes are drawn to novelty in any environment (Sousa, 2011), students behind one using a device will unavoidably find their eyes also drawn to the same device, particularly if the student loads visually rich material on screen, such as a game or video. Because laptops make it even easier to switch tabs and programs than a smartphone, students are especially likely to task-switch when bringing a laptop to class. Even students who manage to resist the temptation to task-switch with laptops are likely to take notes dictation-style, rather than the more learning-aligned methods of summary and synthesis, and thus perform worse than students taking notes by hand (Mueller & Oppenheimer, 2014).

Because instructors are aware that students have difficulty focusing on course content when they do not check their devices often enough, some- times they allow for a "break-for-technology" in the hopes that task-switching would otherwise be kept at bay, and that time spent outside the break would be truly focused on the task at hand (King, 2017). However, when- ever students do use their devices for off-task activities, either by disre- garding the lecture or by sneaking a look at social media after responding to a clicker question, they have a difficult time returning to the primary task, as a majority of them view the smartphone as an irresistible temptation (Witecki & Nonnecke, 2015). The temptation to further use social media is strong at this point, and arguably made stronger by the existence of the break in the first place, which creates a mental connection between private phone usage and the class (King, 2017).

Even students using smartphones while on-task, such as looking up further details and contexts to material discussed in class as we described earlier as a benefit to technology, are using technology in such a way as to endanger their overall path to learning, as least according to the traditional definition of learning. Due to the brain's plasticity, or ongoing ability to rewire itself, our brains become good at what we ask them to do. The ubiquity of smart- phones has enabled learners of all ages, but perhaps especially generations who grew up with internet-enabled phones, to consider facts as information outside of themselves, and only to be looked up when needed (as opposed to internalized as memorized facts). In his landmark book *The Shallows*, Nicholas Carr (2010) argued that easy access to the internet (and search engines in particular) are wiring our brains to become less effective at remembering facts.

Because the very definitions of what constitutes knowledge appear to be changing, or at least the relationship to knowledge, it may be fruitful to consider a taxonomy of possibilities. This proposed taxonomy of proximity to knowledge has four levels:

1. *Distant*: When knowledge is distant, it is viewed as completely outside the learner. Information is accessible on demand, but is not initially present in either short-term or long-term memory. Many students consider dates of historical events to be distant knowledge.

2. *Temporary*: When knowledge is temporary, it is accessible in short-term memory because information has recently been consumed. Cramming before a test (with no other study of the same material) results in this type of temporary memory.

3. *Accessible*: When knowledge is accessible, it's been previously consolidated into long-term memory but needs to be retrieved (technically, reconstructed) into working memory before being usable. Recalling the location of a parked car at a supermarket is an example of accessible memory.

4. *Integrated*: When knowledge is integrated, it is effortlessly and unconsciously available without cognition. Examples might include knowing which kitchen drawer to open to find a fork.

Higher education students may fall into any of these levels of proximity to knowledge. Certainly there will be some who consider college to be of value primarily for the earned credential, and not any of the knowledge (or skills) ostensibly to be learned. In theory, students move down the proximity taxonomy as they become more sophisticated and mature learners. Indeed, parallels could be drawn to William Perry's (1970) scheme of intellectual development, in which he proposed a taxonomy of the complexity of thought processes by learners. At the lowest level, dualistic thinkers see only right and wrong answers. Thinkers at the next level higher, multiplicity, recognize there are multiple potential answers, but choose to trust familiar and instinctive answers as a result. Relativistic thinkers go further in seeing possible truths in all answers and explanations, but are unable to select a single answer. Only those at the highest level, commitment, are able to select and defend a solution to a particular problem, in full recognition of other perspectives and their relative merits. The connection to the proximity to knowledge is at this highest level. A commitment-level thinker is likely to consider knowledge valuable enough to integrate, and not merely access when needed.

It is important to remember, however, that students do not necessarily simply progress through the levels of this taxonomy over time. It would be possible, for instance, for the same student to employ an attitude of integrated knowledge in one class that interests him greatly, but display a distant relationship to knowledge in a required class he finds boring. Thus, the proximity to knowledge is not fixed to a person even at a single point in time, and is more accurately representative of the person's view of the content or context.

Still, it is likely that some students have increasingly come to accept distant and temporary knowledge as sufficient for college-level work, possibly due to the ubiquity of the internet, as Carr (2010) documented. While faculty and students may have different views on the students' optimal proximity to knowledge, it is equally critical to consider employers. Have employers equally changed over time to accept primarily distant and temporary access to knowledge? While this may seem unlikely on the surface, discipline-related knowledge is typically not one of the skills seen as lacking in employer surveys, which instead highlight soft skills such as communication, team-work, and problem solving (Nisha & Rajasekaran, 2018). Additionally, the unceasing march of technology will affect workplaces as well. If interaction with computers becomes primarily verbal, as we've seen for decades in science fiction, then more of society may shift to viewing distant and tempo-rary knowledge as acceptable for a default position.

In my view, college instructors face an immediate problem in the related phenomenon of shortened attention spans. Almost all of the work of college demands sustained attention: reading chapters, completing homework prob-lems, listening to lectures, writing essays, and completing research. With students increasingly less likely to be able to focus for the time periods faculty had in mind when assigning the work, there is a growing mismatch between what faculty want and what students perform (Franke et al., 2010). Indeed, students who use smartphones heavily report finding the device irresistible when attempting to complete homework (Furst et al., 2018). Some instruc-tors try to shift the nature of instruction to match the changing students. One common reaction to the recognition that students don't read textbook chap-ters is to rely on more video for content delivery, a move that also matches students' increasing preference for YouTube over Google when researching common processes or household tasks (Head & Eisenberg, 2010). But this is no panacea. Students increasing want videos to be shorter as well (Slemmons et al., 2018), and at some point the lessened amount of content being deliv-ered will cross a red line in terms of rigor; obviously, there is a different red line for each instructor. Other faculty may accept the need for major semester deliverables to be shorter than ideal, as students increasingly report they lack the time to tackle big projects—a problem likely related to the large percentage of students who hold jobs while attending college, which in 2017 was 43% of full-time students and 81% of part-time students (McFarland et al., 2019). Those faculty who know that attempts at multitasking (in reality, task-switching) by students results in lowered performance on both tasks may try to enforce classroom management rules that disallow personal devices entirely, because students engage in task-switching even when they know it lowers their performance (Mokhtari et al., 2015). Some faculty have accepted the need for information to be distant or temporary, rather than integrated, and have shifted their pedagogy appropriately, primarily by focusing on skills rather than content knowledge (Tulgan, 2015).

In the grand scheme of things, we may see a societal shift in how cognition is valued. If sustained focused attention becomes uncommon, the nature of cognition may be viewed differently. Thinking deeply about a concept may come to mean "overthinking" it. We certainly already see a shift in media toward shorter delivery and shorter content, not only in terms of shorter soundbytes of politicians but also in the proliferation of list-based articles in websites and newspapers with far fewer words overall than traditional articles.

SUGGESTED NEXT STEPS

There are certain steps that institutions of higher education, at the level of both faculty and IT departments, can take to maximize student attention and learning. The first suggestion is to align technology adoption choices with the targeted level of proximity to knowledge. For instance, if a faculty member wishes to prioritize accessible and integrated knowledge—meaning that students would need to memorize and internalize certain facts—then technology can be chosen to aid the pedagogy. Specifically, the instructor can utilize the LMS to mandate distributed and interleaved retrieval practice through the use of cumulative quizzes and quiz banks. IT departments might similarly leverage human factors such as nudges to enhance the student experience. Nudges work to shift behavior by setting certain defaults, such as a doctor's office setting a next appointment at the conclusion of the current one, rather than wait for behavior that may not materialize in the weeks to come. When he was president of Goucher College, José Bowen directed the WiFi access to be intentionally fastest and strongest in the lounge areas of the new residence halls (Biemiller, 2017), thereby nudging students to congregate and stave off loneliness, a known major contributor to lack of college completion (Ponzetti, 1990).

Institutions should be cautious about technology adoption simply to meet popular demand. A prime example can be seen in the installation of WiFi access in lecture halls. While this is student-friendly, it is neutral at best to learning and may be an enabler of poor learning habits. As discussed previously, student-connected devices have only limited targeted use and even those use cases come with caveats such as prolonged distractions.

Even more dangerous is the use of lecture capture. Although it offers benefits to students needing accommodations, in practice it also encourages absenteeism and task-switching in the absence of an authority figure's presence in the same space (Korving et al., 2016), and likely encourages second viewings in favor of spaced retrieval practice (Lambert et al., 2019). At the same time, lecture capture is viewed by students as a popular feature, and discontinuation of an existing practice is likely to be met with protest.

One final technology practice that is popular with students but problematic for the learning process is the distribution of PowerPoint handouts prior to a lecture. Students frequently insist that these handouts be offered, often citing the ability to take "additional" rather than "primary" notes. However, in practice, some students do not take additional notes, relying on the printed notes as sufficient (Witherby & Tauber, 2019). With notes already in hand, it seems likely to me that students' attention is less likely to be focused on the intricacies of the lecture, making attention-wandering possible, and even when attention remains on the lecture, enhanced passivity endangers the possibilities for deep learning.

The first suggestion is to align technology adoption choices with the targeted level of proximity to knowledge. Faculty and support offices with expertise in student learning should have a voice in deciding which technologies are implemented. As faculty are on the front lines of learning, they will have insights into likely student behavior given a particular new technology, and their instincts are likely to form a valuable prediction of effects. Another best practice for campus technology is to incorporate faculty as deeply as possible in the decision-making process for technology adoption.

While this discussion has addressed many positive and negative effects of technology on attention in the college environment, it is neither comprehensive nor conclusive. Further work needs to be done, for instance, on student cognitive load limitations with regard to personalized learning environments. We know that students do not know every technology, and that the storyline of them being "digital natives" with innate tech savvy is largely false (Margaryan et al., 2011), so they are likely to need instruction on educational technologies. As personalized learning environments broaden, so too does the proliferation of available technologies from which to customize, and cognitive load limits will be a concern.

Another area ripe for investigation is the effect of student metacognition on the limitations of technology. Will students take effective additional notes on PowerPoint handouts if given instruction on the science of learning? Can students use laptops for synthesis note-taking, rather than the less-useful dictation, if they are given enough information to make metacognitive connections to their learning? I think we are likely to find that metacognition enhances learning sufficiently to overcome the temptations that some technologies can present, but additional research will be required.

A final note about the attention economy and higher education: faculty and instructors have not been immune to the effects of the shifts in society, and our own attention spans may have dwindled accordingly. Indeed, our seemingly insatiable appetite to discover the next paradigm shift in learning, through technology or otherwise, bespeaks our inability to take the long view on tech affordances, their promise, and the true cost of learning.

REFERENCES

Alexander, C. (2018, August 10). *Microsoft Office Mix [RIP]: Why was it offered at all?* https://elearningindustry.com/microsoft-office-mix-rip-offered

Alibec, C., & Sandiuc, C. (2015, April 23–24). *The benefits of e-learning through webquests. A case study* [Conference session]. The 11th International Scientific Conference "eLearning and Software for Education," Bucharest, Romania.

Au, W. J. (2013, June 23). Second Life turns 10: What it did wrong and why it may have its own second life. *Gigaom.* https://gigaom.com/2013/06/23/second-life-turns-10-what-it-did-wrong-and-why-it-will-have-its-own-second-life/

Bell, K. (2017). *Game On!: Gamification, gameful design, and the rise of the gamer educator.* Johns Hopkins University Press.

Biemiller, L. (2017, August 4). *What's new in freshman housing? Buildings that help students make friends.* https://www.chronicle.com/article/What-s-New-in-Freshman/240864

Bowman, N. D. (2018). *Video games: A medium that demands our attention.* Routledge.

Brown, P. C., Roediger, H. L., III, & McDaniel, M. A. (2014). *Make it stick: The science of successful learning.* Belknap Press. https://doi.org/10.4159/9780674419377

Burge, K. (2018, December 27). Virtual reality helps hospice workers see life and death through a patient's eyes. *NPR.* https://www.npr.org/sections/health-shots/2018/12/27/675377939/enter-title

Canali, R. (2016, March 10). *Game design for eco driving.* http://www.megamification.com/game-design-for-eco-driving/

Carr, N. (2010). *The Shallows.* W. W. Norton.

Chen, X., & Han, T. (2019). Disruptive technology forecasting based on Gartner Hype Cycle. *IEEE.* https://doi.org/10.1109/TEMSCON.2019.8813649

Chin, K., Kao, Y., & Hsieh, H. (2018, July 8–13). *A virtual reality learning system to support situational interest in historic site courses* [Conference session]. 7th International Congress on Advanced Applied Informatics. Yonago, Tottori, Japan.

Christensen, C. M., Horn, M. B., & Johnson, C. W. (2010). *Disrupting class: How disruptive innovation will change the way the world learns.* McGraw Hill.

Deci, E. L. (1975). *Intrinsic motivation.* Springer. https://doi.org/10.1007/978-1-4613-4446-9

Deci, E. L., & Ryan, R. M. (1985). *Intrinsic motivation and self-determination in human behavior.* Springer. https://doi.org/10.1007/978-1-4899-2271-7

Deci, E. L., & Ryan, R. M. (2002). Overview of self-determination theory: An organismic dialectial perspective. In R. M. Ryan & E. L. Deci (Eds.), *Handbook of self-determination research* (pp. 3–33). University of Rochester Press.

Deterding, S., Khaled, R., Nacke, L., & Dixon, D. (2011). *Gamification: Toward a definition.* Proceedings of CHI 2011 Workshop, Gamification: Using game design elements in non-game contexts, May 7–12, Vancouver, BC, Canada.

Dunlosky, J., Rawson, K. A., Marsh, E. J., Nathan, M. J., & Willingham, D. T. (2013). Improving students' learning with effective learning techniques: Promising directions from cognitive and educational psychology. *Psychological Science in the Public Interest, 14*, 4–58. https://doi.org/10.1177/1529100612453266

Franke, R., Ruiz, S., Sharkness, J., DeAngelo, L., & Pryor, J. (2010). *Findings from the 2009 Administration of the College Senior Survey (CSS): National aggregates.* University of California, Los Angeles, Higher Education Research Institute.

Freeman, S., Eddy, S. L., McDonough, M., Smith, M. K., Okoroafor, N., Jordt, H., & Wenderoth, M. P. (2014). Active learning increases student performance in science, engineering, and mathematics. *Proceedings of the National Academy of Sciences, 111*(237), 8410–8415.

Frischmann, B. (2019, September 20). There's nothing wrong with being a Luddite. *Scientific American Mind, 30*, 1.

Furst, R. T., Evans, D., & Roderick, N. M. (2018). Frequency of college student smartphone use: Impact on classroom homework assignments. *Journal of Technology in Behavioral Science, 3*(2), 49–57.

Gabana, D., Tokarchuk, L., Hannon, E., & Gunes, H. (2017, October 23–26). *Effects of valence and arousal on working memory performance in virtual reality gaming* [Conference session]. 7th International Conference on Affective Computing and Intelligent Interaction, San Antonio, TX.

Grant, P., & Basye, D. (2014). *Personalized learning: A guide for engaging students with technology.* International Society for Techology in Education.

Handoko, E., Gronseth, S., McNeil, S., Bonk, C., & Robin, B. (2019). Goal setting and MOOC completion: A study on the role of self-regulated learning in student performance in massive open online courses. *International Review of Research in Open and Distributed Learning, 20*(3), 39–58. https://doi.org/10.19173/irrodl.v20i4.4270

Head, A. J., & Eisenberg, M. B. (2010). *Truth be told: How college students evaluate and use information in the digital age.* Project Information Literacy.

Heath, C., & Heath, D. (2007). *Made to stick: Why some ideas survive and others die.* Random House.

Hoffman, M., & Omsted, R. (2018). Credentials for open learning: Scalability and validity. *International Journal of Open Educational Resources, 1*(1). https://doi.org/10.18278/ijoer.1.1.4

Hubbard, J., & Couch, B. (2018). The positive effect of in-class clicker questions on later exams depends on initial student performance level but not question format. *Computers & Education, 120*, 1–12. https://doi.org/10.1016/j.compedu.2018.01.008

Imazeki, J. (2014). Bring-your-own-device: Turning cell phones into forces for good. *The Journal of Economic Education, 45*(3), 240. https://doi.org/10.1080/00220485.2014.917898

Judd, T., & Kennedy, G. (2011). Measurement and evidence of computer-based task switching and multitasking by 'Net Generation' students. *Computers & Education, 56*(3), 625–631. https://doi.org/10.1016/j.compedu.2010.10.004

King, B. (2017, October 19). Should college professors give 'tech breaks' in class? *NPR.* https://www.npr.org/sections/13.7/2017/10/19/558751178/should-college-professors-give-tech-breaks-in-class

Korving, H., Hernandez, M., & DeGroot, E. (2016). Look at me and pay attention! A study on the relation between visibility and attention in weblectures. *Computers & Education, 94*, 151–161. https://doi.org/10.1016/j.compedu.2015.11.011

Kuznekoff, J., & Titsworth, S. (2013). The impact of mobile phone usage on student learning. *Communication Education, 62*(3), 233–252. https://doi.org/10.1080/03634523.2013.767917

Kyriazi, T. (2015). Using technology to introduce frequent assessments for effective learning: Registering student perceptions. *Procedia: Social and Behavioral Sciences, 197*, 570–576. https://doi.org/10.1016/j.sbspro.2015.07.195

Lambert, S., Pond, K., & Witthaus, G. (2019). Making a difference with lecture capture? Providing evidence for research-informed policy. *International Journal of Management Education, 17*(3), 100–323. https://doi.org/10.1016/j.ijme.2019.100323

Lepp, A., Barkley, J. E., Karpinski, A. C., & Singh, S. (2019). College students' multitasking behavior in online versus face-to-face courses. *SAGE Open, 9*(1), 1–9. https://doi.org/10.1177/2158244018824505

Margaryan, A., Littlejohn, A., & Vojt, G. (2011). Are digital natives a myth or reality? University students' use of digital technologies. *Computers & Education, 56*(2), 429–440. https://doi.org/10.1016/j.compedu.2010.09.004

McFarland, J., Hussar, B., & Zhang, J. (2019). *The condition of education 2019*. National Center for Education Statistics.

Merchant, Z., Goetz, E., Cifuentes, L., Kenney-Kennicutt, W., & Davis, T. (2014). Effectiveness of virtual reality-based instruction on students' learning outcomes in K–12 and higher education: A meta-analysis. *Computers & Education, 70*(1), 29–40. https://doi.org/10.1016/j.compedu.2013.07.033

Mokhtari, K., Delello, J., & Reichard, C. (2015). Connected yet distracted: Multitasking among college students. *Journal of College Reading & Learning*. https://doi.org/10.1080/10790195.2015.1021880.

Mueller, P. A., & Oppenheimer, D. M. (2014). The pen is mightier than the keyboard: Advantages of longhand over laptop note taking. *Psychological Science, 25*(6), 1159–1168. https://doi.org/10.1177/0956797614524581

Nisha, M., & Rajasekaran, V. (2018). Employability skills: A review. *The IUP Journal of Soft Skills, 12*(1), 29–37.

O'Donnell, S., & Epstein, L. H. (2019). Smartphones are more reinforcing than food for students. *Addictive Behaviors, 90*, 124–133. https://doi.org/10.1016/j.addbeh.2018.10.018

Owston, R., Lupshenyuk, D., & Wideman, H. (2011). Lecture capture in large undergraduate classes: Student perceptions and academic performance. *The internet and Higher Education, 14*(4), 262–268. https://doi.org/10.1016/j.iheduc.2011.05.006

Perry, W. G. (1970). *Forms of intellectual and ethical development in the college years: A scheme*. Holt, Rinehart, and Winston.

Pertanika J. (2018). Enhancing active learning in large classes using web clicker. *Journal of Social Sciences and Humanities, 26*, 111–120.

Ponzetti, J. J. (1990). Loneliness among college students. *Family Relations, 39*(3), 336–340. https://doi.org/10.2307/584881

Rana, N. P., Dwivedi, Y. K., & Al-Khowaiter, W. A. (2016). A review of literature on the use of clickers in the business and management discipline. *International*

Journal of Management Education, 14(2), 74–91. https://doi.org/10.1016/j.ijme.2016.02.002

Rosen, L. D. (2012, December 18). Weapons of mass distraction: Why we have lost the ability to focus. https://www.psychologytoday.com/us/blog/rewired-the-psychology-technology/201212/weapons-mass-distraction

Sahlström, F., Tanner, M., & Valasmo, V. (2019). Connected youth, connected classrooms. Smartphone use and student and teacher participation during plenary teaching. *Learning, Culture and Social Interaction, 21*, 311–331. https://doi.org/10.1016/j.lcsi.2019.03.008

Scolari, C. A., & Contreras-Espinosa, R. S. (2019). How do teens learn to play video games? Informal learning strategies and video game literacy. *Journal of Information Literacy, 13*(1), 1.

Sears, R., Hopf, F., Torres-Ayala, A., Williams, C., & Skrzypek, L. L. (2019). Using plan-do-study-act cycles and interdisciplinary conversations to transform introductory mathematics courses. *PRIMUS, 29*(8), 881–902. https://doi.org/10.1080/10511970.2018.1532938

Slemmons, K., Anyanwu, K., Hames, J., Grabski, D., Mlsna, J., Simkins, E., & Cook, P. (2018). The impact of video length on learning in a middle-level flipped science setting: Implications for diversity inclusion. *Journal of Science Education and Technology, 27*(5), 469–479. https://doi.org/10.1007/s10956-018-9736-2

Sousa, D. A. (2011). *How the brain learns* (4th ed.). Corwin.

Stowell, J. R. (2015). Use of clickers vs. mobile devices for classroom polling. *Computers & Education, 82*, 329–334.

Sultan, N., & Jamal Al-Lail, H. (Eds.). (2015). *Creative learning and MOOCs: Harnessing the technology for a 21st century education*. Cambridge Scholars.

Sunstein, C. R., Reisch, L. A., & Kaiser, M. (2019). Trusting nudges? Lessons from an international survey. *Journal of European Public Policy, 26*(10), 1417–1443. https://doi.org/10.1080/13501763.2018.1531912

Tulgan, B. (2015). *Bridging the soft skills gap: How to teach the missing basics to today's young talent*. Jossey-Bass. https://doi.org/10.1002/9781119171409

Varnavsky, A. N. (2016). *Research of preference in playback speed of learning video material depending on indicators of cognitive processes*, pp. 1–4. Cognitive Sciences, Genomics and Bioinformatics (CSGB), Novosibirsk, Russia. https://doi.org/10.1109/CSGB.2016.7587686

Wang, W., & Wang, C. (2009). An empirical study of instructor adoption of web-based learning systems. *Computers & Education, 53*(3), 761–774. https://doi.org/10.1016/j.compedu.2009.02.021

Watt, S., Vajoczki, S., Voros, G., Vine, M., & Fenton, N. (2014). Lecture capture: An effective tool for universal instructional design? *Canadian Journal of Higher Education, 44*(2), 1–29.

Willingham, D. (2009). *Why don't students like school? A cognitive scientist answers questions about how the mind works and what it means for the classroom*. Jossey-Bass.

Witecki, G., & Nonnecke, B. (2015). Engagement in digital lecture halls: A study of student course engagement and mobile device use during lecture. *Journal of Information Technology Education, 14*, 73–90. https://doi.org/10.28945/2103

Witherby, A. E., & Tauber, S. K. (2019). The current status of students' note-taking: Why and how do students take notes? *Journal of Applied Research in Memory and Cognition, 8*(2), 139–153. https://doi.org/10.1016/j.jarmac.2019.04.002

5

"SAY CHEESE!"

How Taking and Viewing Photos Can Shape Memory and Cognition

LINDA A. HENKEL, ROBERT A. NASH, AND JUSTIN A. PATON

Many of us regularly get caught up in documenting our personal experiences by taking copious photos and videos and then sharing them via social media websites. But because our attentional resources are so stretched by competing demands at any given moment, it is unsurprising that we sometimes fail to be fully present or engaged in the very experiences we are documenting. Taking photos can, in some cases, help people to attend to their surroundings; yet at other times, they can be guilty of treating photos merely as trophies to collect. For example, while walking through a museum and after waiting patiently to approach a great work of art, people glance upward at the wall momentarily, hastily snap their cameras, and say, "What's next?" They hike to a beautiful waterfall and, on reaching the top, pose for a selfie, rapidly shifting their attention to the number of "likes" their newly shared photo is earning. They even attend concerts, plays, and sporting events, and watch the small screens in their hands while video-recording their experiences rather than watching the live action happening right in front of them.

This chapter explores the interplay between attention and memory in the context of taking and viewing photos. What we pay attention to, how much

https://doi.org/10.1037/0000208-006
Human Capacity in the Attention Economy, S. Lane and P. Atchley (Editors)

attention we devote, and the kind of attention we engage can all influence our subsequent memory for our experiences. People rely on photos as memory cues as a means to mitigate the cognitive limitations in encoding, storing, and retrieving their experiences. But in many ways, these pronounced limitations are all the more severe with the increasing availability of modern technology that allows people to simultaneously check email, send messages, and update their social media feeds while going about their everyday lives. This chapter discusses research that suggests that because of these limitations in attentional and cognitive resources, people sometimes treat cameras as an external memory device—in essence, offloading their memories onto the camera. We explore both the positive and negative consequences for memory of taking and viewing photos by outlining studies of what happens when we divide our attention between experiencing events and documenting them with a camera, and outlining studies on the attentional and memory demands created by the ensuing accumulation of photos.

HOW MANY PHOTOS ARE PEOPLE TAKING, AND WHY?

In the almost 200-year history of photography, changes in technology and in societal norms have dramatically shaped the ways in which people take and use photos. Selfies and plates of food may currently be all the rage in terms of what people like to capture, but, of course, this was not always the case. The first ever photograph was taken in 1826, and the first one with a person featured in the shot was taken in 1838. Commercial cameras for the general public were not introduced until 1900, and by 1930, approximately a billion photos were being taken worldwide each year (Good, 2011). Photo-taking continued to grow steadily in popularity throughout the remainder of the 20th century. Commercial digital cameras were introduced to the public in 1994, and as technological advances continued with the first digital camera phone marketed in 1997, digital photography became more affordable and accessible to the public, soaring in popularity. It has been estimated that 86 billion photos were taken in the year 2000 alone, 300 billion in 2010, and in 2017, a projected 1.3 trillion per year (Good, 2011; Heyman, 2015).

The exponential growth of photo-taking is staggering, going hand-in-hand with growth in the accessibility and usage of the internet, smartphones, social media websites, photo-sharing apps, and inexpensive storage. An estimated 3.5 billion photos and videos are shared each day on average on the social media app Snapchat (C. Smith, 2020). Likewise, in recent years, Instagram users shared an estimated 52 million photos each day ("Instagram Company Statistics," 2016), and an estimated 300 million photos were uploaded to Facebook each day (Noyes, 2020). Facebook reportedly has a photo collection numbering more than 140 billion, or 10 times larger than the Library of Congress in its entirety (Meeker, 2014). When all sites and platforms are combined, in 2014, it was estimated that approximately 1.8 billion photos were uploaded and shared each day, equivalent to 657 billion per year (Meeker, 2014). By 2015, these numbers had soared to 3.2 billion photos uploaded each day (Meeker, 2016).

But what are all these photos for? People use photos as a way of documenting and recording special occasions and everyday moments, and many people report that they use photography primarily as a tool for helping them remember these experiences (Chalfen, 1998; Harrison, 2002; Whittaker et al., 2010). Recent surveys have shown that although the intent to preserve memory cues still remains a primary reason for taking photos—with 37% of people reporting this as their main reason—a further 36% of people reported a main objective of documenting their experiences by "capturing the moment as it is" (Shutterfly, 2014). Instagram and Facebook posts can have a "look at me, look at what I am doing" quality, and, in this sense, photos are instrumental in identity formation and presentation, and are used as a means for people to communicate with each other (van Dijck, 2008). Indeed, in another recent survey, a solid majority of people reported that they take photos to "capture memories for themselves" (59%), but a large percentage (28%) also said their primary goal was "to share with others" (Diehl et al., 2016). In a follow-up survey, respondents were asked to indicate what percentage of the photos they took were for each of a specified list of purposes ("for my own memory," "to share with others," "another goal," and "no particular goal"). Results again showed that the majority of photos people took were intended for aiding their own memory (58%), and a solid one third (32%) of photos were taken for the purpose of sharing with others. In addition to cueing our memories and communicating with others, photos can also serve as a source of evidence of what really happened. Research shows that when trying to verify what really happened in the past, people prefer verification strategies that are reliable, but more crucially, they prefer strategies that are cheap and easy to use; photos increasingly serve both of these important goals (Nash et al., 2017; Wade et al., 2014).

Whatever their function, people do value their photos. When given the hypothetical option of either saving their cell phone with no photos on it or saving the photos themselves but losing the phone, 71% of people surveyed said they would rather save the photos (Shutterfly, 2014). Given the high importance people place on their photos, one might expect that they would go to great lengths to organize and securely store them. But despite their best intentions, people report being overwhelmed by the number of photos they accumulate and by the formidable task of organizing these unwieldy collections. In turn, this sense of being overwhelmed discourages many people from spending time actually looking at their photos and reminiscing (Bowen & Petrelli, 2011; Ceroni, 2018; Nunes et al., 2009; Shutterfly, 2014). In their discussion of the "attention economy," Atchley and Lane (2014) noted Herbert Simon's observation that "a wealth of information creates a poverty of attention" (p. 136), and that sentiment is reflected clearly in people's photo-curating behavior. In a survey of people who regularly used their smartphones for taking photos, the majority (74%) reported that they almost never or just occasionally sort through the photos on their devices (Hartman, 2014). More than half (57%) believed their collections were largely disorganized or somewhat organized but needed a lot of work, and 40% reported they did not have the discipline to more successfully organize their photos (Hartman, 2014).

In a study that brings this issue to light, parents with young children were tasked with finding a series of specific, valued photos that they had taken more than a year prior (Whittaker et al., 2010). This group of parents reported being "active curators" of photos, which they accumulated for the primary purpose of later facilitating their own and their children's memories. Consequently, many of the parents were confident that they would succeed at the photo-finding task. Yet, they notably overestimated their abilities, failing to find the sought after photo on their computers or other storage devices on nearly 40% of the trials. Reflecting on how challenging the task was in contrast to their expectations, participants reported that they took too many pictures, stored different photos on different devices, and had minimal and ineffective hierarchical organization (e.g., folders organized by date or by overly broad labels).

VIEWING PHOTOS CAN POSITIVELY IMPACT WHAT PEOPLE REMEMBER

Given the large number of photos people take and that a major aim of taking photos is to review them later on and cue memories, it is important to ask what impact viewing photos actually has on our attention and in turn on our memory. In several studies, for instance, after people reviewed photos of events they had witnessed, they were subsequently able to recall more details than were people who had not reviewed photos; this was true both for younger and healthy community-dwelling older adults (Koutstaal et al., 1998; Schacter et al., 1997). In addition, older adults experiencing memory impairments were often found to benefit from the rich retrieval cues afforded by looking at photos (Bourgeois et al., 2001; Heyden, 2014). For example, when people with dementia look at personal photos (e.g., of themselves, family, friends), it draws and engages their attention more than viewing other visual stimuli (Yasuda et al., 2009). Reviewing photos in this way has been reported to promote positive mood and increased feelings of well-being in people with dementia or mild cognitive impairment (Crete-Nishihata et al., 2012), and it has been used effectively within structured reminiscence therapy to improve the mood, well-being, and cognitive functioning of people with dementia (Subramaniam et al., 2014).

Young children, too, can benefit from the use of photos as retrieval aids. For example, when children—even those as young as 2 years old—review photos of their experiences, they subsequently remember more correct details about the experiences (Aschermann et al., 1998; Deocampo & Hudson, 2003; Jack et al., 2015). Researchers in one study examined 5- and 6-year-olds' memory of a class field trip to a museum in which the teacher took photos of the children's different activities, such as walking from their school to the museum, digging in a sandbox for archaeological artifacts, and making clay models (Fivush et al., 1984). Memory for the trip was assessed in interviews that occurred 6 weeks and 1 year later. In these interviews, the children were shown the trip photos, and they indicated whether they remembered what was happening in each one; they then had to indicate which event occurred first, second, third, and so forth. Six weeks after the museum visit, the children reported remembering almost all of the events depicted in the photos (they recognized 5.6 out of 6 photos on average). Their temporal memory was also rather impressive: The majority of children arranged all of the photos in correct temporal order. One year after the museum visit, children still reported recognizing almost all of the photos and ordered them in temporal sequence just as accurately as they had at the 6-week delay. Six years later, when the same children were 11–12 years old, a follow-up study was conducted that demonstrates the benefits of photo review (Hudson & Fivush, 1991). In general, when given no photo cues at this time point, the children required specific verbal prompting to recall the museum visit, and, understandably, they recalled less information about the visit than they had 6 years prior. However, when the photos were provided as retrieval cues, these helped them to remember: The photos prompted them to recall roughly the same amount of information about the trip after this 6-year gap as they had remembered when viewing the photos just 6 weeks after the trip. These findings illustrate the power of photos as retrieval cues for young children, whose memory abilities are still rapidly developing (see also Salmon, 2001).

As in these studies with children, several other studies have used the naturalistic setting of museums as a context for examining how reviewing photos of past experiences can shape memories across the life span. In some of these studies, for example, young adults took photos while on a self-guided museum tour (St. Jacques et al., 2013; St. Jacques & Schacter, 2013). To prevent them being distracted by the act of taking photos, participants wore a camera on a strap around their neck that automatically captured what they were facing every 15 seconds. Two days later, they reviewed a subset of the photos taken by the automatic camera from some of the exhibits on the tour, and these were intermixed with photos from a different tour that they had not been on. Another 2 days later, the experimenters assessed what participants remembered about the tour by showing them various new and old photos and asking which ones depicted scenes from their actual tour. The results showed that when people had earlier reviewed photos of parts of their museum visit, this photo review enhanced their memory of their experiences and produced a stronger feeling of "reliving" the moment. Moreover, the more strongly they felt a sense of reliving while reviewing the photos, the greater the boost in their subsequent retention of those experiences. These findings were replicated in a later study involving both younger and older adults, and although both age groups showed memory benefits from photo review, the relative boost in memory performance was more modest for older adults (St. Jacques et al., 2015).

In another study, people took a guided tour of a museum while either not taking any photos, taking photos of some of the artwork, or posing next to selected artwork as the tour guide took photos of them (Parisi & Henkel, 2014). Three days later, the participants either reviewed the photos that were taken or they reviewed only the names of the artwork. Then, 1 week after reviewing the photos or names, their memory for the museum tour was assessed. Results showed that reviewing photos of the museum tour increased people's memory for visual details about the artwork they had seen in the exhibit relative to when photos were not reviewed. However, memory for visual details increased only for the specific artwork for which photos were both taken and reviewed. And although reviewing photos boosted memory relative to not reviewing photos, memory accuracy was not influenced by the type of photos reviewed: People's memory was boosted to the same extent regardless of whether they had taken the photo of the artwork themselves or were featured in the shot standing next to the artwork.

A key question of interest in that study was whether reviewing photos influenced the perspective people experienced when remembering the museum tour: Did they remember the exhibit as if they were seeing it through their own eyes (i.e., a field perspective) or as if they were an observer seeing themselves in the scene (i.e., an observer perspective)? The data showed that after having reviewed photos, people reported a stronger observer perspective in their memories compared with having not reviewed photos (Parisi & Henkel, 2014). This was the case regardless of whether the photos showed the participant standing next to the artwork or just showed the artwork. Thus, photo review influenced the phenomenal characteristics of people's memories.

Technology has made it possible to capture more of our lives in photos than many people may have ever thought possible. This point is illustrated particularly well by passive camera systems, such as SenseCam®, Vicon Revue®, and Autographer®, which people wear as they go about their daily activities. These camera systems take photos automatically throughout the day, and the individual can then review the day's photos in rapid sequence as a "minimovie" soon thereafter. In principle, the automatic nature of these cameras greatly reduces attentional demands brought on by competition between attending to the camera at the expense of attending to the experience and subsequently impacts memory. Studies have shown that both younger and older adults experience benefits in memory from reviewing the day's photos in this way: Retention of detail increases and there is a stronger sense of reliving the experiences (Barnard et al., 2011; Finley et al., 2011; Mair et al., 2017; Seamon et al., 2014; St. Jacques et al., 2011). Case studies and experimental studies have also examined the impact of wearable passive camera systems as external memory aids for clinical populations, such as people with Alzheimer's disease, amnesia, and other memory impairments. This work has likewise found memory benefits arising from photo review (Brindley et al., 2011; Browne et al., 2011; Hodges et al., 2011; Loveday & Conway, 2011; Pauly-Takacs et al., 2011; Woodberry et al., 2015). Indeed, in a recent review of research on wearable passive cameras, the authors noted that the memory benefits of this kind of photo review are especially pronounced in people with memory impairments (Chow & Rissman, 2017).

WHAT IS IT ABOUT VIEWING PHOTOS THAT HELPS PEOPLE REMEMBER?

The findings described so far converge to suggest that reviewing photos of our experiences can provide important benefits to remembering. As captured in the saying "A picture is worth a thousand words," photos can supply vivid visual detail about the people, objects, locations, and events depicted in them. The richness and specificity of photos draws attention and provides concrete and distinctive retrieval cues that can help people reactivate and remember their experiences. For instance, suppose you were looking at a photo of a birthday party from your childhood, which showed you in your special outfit for the day while standing in front of the swing set at the playground. You might be flooded with memories such as, "Oh, I remember those red shoes!" or "The party was at the park down the street from my house, and the weather was perfect." But past events are also organized in memory in an associative way such that similar events and related details are represented in ways that allow them to trigger one another effortlessly (Collins & Loftus, 1975). For this reason, the memories triggered by looking at a photo are not always limited to just the visual information contained within the image: People can remember associated details, their feelings and motivations, and contextual details not depicted in the photo. For example, you might remember "I wanted those red shoes for so long, and a week before my birthday, my mom finally bought them for me as surprise when we were at the mall." Or, you might remember a detail from the party itself but that is not depicted in the photo: "That was the party where my brother fell and cut his knee, and had to get stitches. We were stuck in the emergency room for hours."

Photos can easily grab our attention, and they provide rich and distinctive visual details that gives them a memory advantage as seen in the *picture superiority effect*, whereby people are more likely to remember pictures as compared with words or verbal descriptions (Madigan, 1974; Nelson et al., 1976; Paivio et al., 1968). People's ability to recognize photos they have seen before can be remarkably good, even when the photos are nonpersonal (Shepard, 1967) and even when the participants are young children (Fivush et al., 1984; Pathman et al., 2013). Personal photos are most likely to be effective as retrieval cues when they contain unique and distinctive information that has minimal overlap with other experiences, such as information about the specific activity, rather than more general information, such as a frequently visited location (Burt et al., 1995). The memory advantage for one's own photos is seen in a study in which people reviewed photos they had taken while walking around a college campus (Cabeza et al., 2004). They showed different patterns of brain activation compared with when they viewed similar photos taken by other people. According to these patterns of brain activation, people engaged more self-referential processing, visual–spatial memory, and recollective experience than when they viewed their own rather than other people's photos.

Insight into how photos impact the interplay between attention and memory also comes from those studies that use wearable cameras. The memory benefits of reviewing the daily minimovies from these cameras arise in part because the act of reviewing photos is, in essence, a second exposure to the events. Several of the aforementioned studies used a control condition that involved reviewing diary-type verbal descriptions, and these studies found that photo review boosted memory in terms of both quantity and quality relative to reviewing these verbal descriptions (Hodges et al., 2011). In addition, photo review can prompt people to engage in additional reflection and cognitive processing, which produces positive side effects such as improving mood and feelings of well-being (Murphy et al., 2011) as well as enhanced performance on neuropsychological tests assessing memory and executive functions (Silva, Pinho, Macedo, & Moulin, 2013). In reviewing 25 research studies on the effects of wearable cameras on memory, Silva et al. (2018) noted that many of the studies focused solely on memory for details seen in the photos and that few studies have looked at memory for what they called "something more" (p. 121). They purported that some limited evidence suggests that reviewing images from wearable cameras can improve memory beyond the information contained in the photos by reinstating details, such as one's prior thoughts and feelings, that in turn triggers additional nondepicted information. Silva et al. (2018) hypothesized that because of the first-person perspective of the automatically taken photos, the large volume of photos taken, and the manner in which they are viewed as a minimovie, reviewing the images should "act as a powerful cue to raise the activation of the event above threshold . . . [making] information and detail available to the experiment, even if it is not represented in the image" (p. 123). The authors noted the need for more theory-driven research to examine this possibility.

A DOUBLE-EDGED SWORD: VIEWING PHOTOS CAN ALSO NEGATIVELY IMPACT MEMORY AND COGNITION

On the basis of the work just cited, it would be easy to conclude that our penchant for capturing our diverse experiences—from the momentous to the mundane—provides us a highly effective way of avoiding the memory flaws that our limited attentional resources can cause. Yet, it is also important to consider how a photo review can be a double-edged sword. In our everyday lives, just as we are selective in what we opt to take photos of, we are also selective in terms of which photos we choose to look at later those we choose to not look at. For instance, a person who has gone through a bitter divorce or has lost a loved one might avoid looking at photos of their wedding or their life together. Such attentional selectivity has consequences for subsequent memory, as illustrated in a study in which participants performed a series of unusual activities (e.g., tracing a boomerang, hammering a nail into a woodblock) and then later either looked at photos of a selection of those activities or did not (Koutstaal et al., 1999). Results showed that people who reviewed photos of completed activities later recalled more of those activities than did those who saw no photos. However, the photo review group actually demonstrated poorer memory for those activities that they did not review in photos compared with people who saw no photos at all. Reviewing photos improved memory for the reviewed experiences, but it impaired memory for the nonreviewed experiences.

Beyond the costs caused by reviewing some photos but not others, there are even more striking consequences in that viewing photos can bring about false beliefs and memories of what happened. Decades of research have shown that espousing false beliefs and memories can create minor issues and inconveniences in one's life (e.g., when you swear you brought your umbrella but you did not; falsely remembering a trivial event from childhood that you did not in fact experience), but it can also create major issues that have significant consequences (e.g., falsely remembering details of a crime when testifying in court; creating intense friction in family dynamics when arguing about what "really happened"; incorrectly remembering that you have already taken your medicine; e.g., Johnson et al., 2011). Research has examined the consequences of viewing photos in a wide range of settings, including readers' interpretations of stories they read in the media, eyewitnesses' recollections of crimes, people's memory for their own actions, and people's beliefs about their own recent and childhood experiences. For example, after reading a newspaper article about a hurricane that hit a town, people were more likely to falsely remember reading about injuries and deaths if they saw a photo depicting the town after rather than before the storm (Garry et al., 2007). After witnessing a mock crime, witnesses were subsequently more likely to identify an innocent suspect from a lineup if, in the interim, that innocent suspect had already been seen in mugshot photos (Deffenbacher et al., 2006; see also Morgan et al., 2013). In other studies, people witnessed or directly experienced events, such as a picnic, and then later saw photos ostensibly from those events. Some of the photos were actions and objects that were part of the events, and some were plausible but did not actually occur (e.g., a photo of a wine bottle on a picnic blanket; Koutstaal et al., 1998, 1999; Schacter et al., 1997). Results revealed that when people reviewed photos of objects and events that were not actually seen, they often claimed that they did in fact witness or experience them as part of the original events.

Other work has looked at how viewing photos can even create false beliefs and memories for actions that people completed themselves. In one line of research, people performed a series of actions (e.g., rip the paper, pour the water into the glass) and then later they saw photos that presented apparent evidence that the actions had been completed (e.g., a photo of torn paper or a full cup of water; Henkel, 2011; Henkel & Carbuto, 2008; Ianì et al., 2018). Some of these photos were new, depicting actions they had not truly done (e.g., a photo of a broken toothpick, an opened piece of candy). A week later, when asked to remember which actions they had actually performed in the first session, people were more likely to falsely claim they remembered performing actions if, in the interim, they had seen a photo of the completed action. Moreover, the rate of these false claims became greater as the number of times they viewed the suggestive photos increased. This *photo inflation effect* shows that photographic "evidence" can make people begin to remember performing actions that they did not in fact perform. Along similar lines, observing other people performing simple actions in video clips can increase participants' false beliefs and memories about having actually performed those actions themselves (Lindner et al., 2010; see also Nash, Wade, & Brewer, 2009).

Viewing photos can even lead us to create false beliefs about visiting locations we never visited. After viewing photos of novel locations, such as college campuses they had not been to, people experienced déjà vu-like feelings of familiarity and were more likely to erroneously claim they had really been to those locations (Brown & Marsh, 2008). Other work in the naturalistic setting of a museum has shown that although people generally remember more and report more vivid memories if they had earlier reviewed photos of their museum trip, the reactivation induced by reviewing the photos also created a breeding ground for false memories (St. Jacques et al., 2013, 2015; St. Jacques & Schacter, 2013). Specifically, when people were exposed to novel, false information about the museum trip while they reviewed their trip photos, they often incorporated that novel information into their subsequent recollections of the trip.

Photos can also bring about false beliefs and memories of more distant childhood experiences, as seen in a study in which college students each reviewed a class photo from their elementary school days while trying to remember a specific event from that school year: a prank in which a classmate had put slime in the teacher's drawer (Lindsay et al., 2004). This slime event was fictional—it did not really happen—but nevertheless many people seemed to remember that it did happen. More important, those students who got to review their old class photo were substantially more likely to falsely report memories of experiencing the event. The class photo, of course, contained no evidence of the slime incident occurring, yet the authors proposed that just looking at the photo could have triggered memories of people, places, and events from that time period. These related memories could then easily be imported into an imagination of pulling a prank on the teacher, enriching the imagination in ways that would make it feel much more like a real memory. Along similar lines, when people look at generic photos (e.g., a photo of a beach or pool scene) while thinking about and envisioning a possible related childhood experience (e.g., being rescued from the water while swimming), they can become more confident that they did indeed experience that event regardless of whether it really happened or not (Blandón-Gitlin & Gerkens, 2010; see also Braun et al., 2002; Hessen-Kayfitz et al., 2017). The consequences of these erroneous beliefs and memories can range from relatively inconsequential to ones with significant impact on people's own lives and the lives of others.

WHY MIGHT VIEWING PHOTOS LEAD TO DISTORTED BELIEFS AND MEMORIES?

People often use photo-taking as a way of preserving the past as it truly happened. But, viewing photos can apparently distort people's reconstructions of the past, which is especially surprising (Wade et al., 2017). What might explain the power of photos to alter the past in this way?

One obvious possibility is that because photos are typically a highly credible source of information about the past, people find it easy to believe that everything depicted in a photo truly happened, which in turn makes it easier to believe that they actually remember those things occurring (Mazzoni & Kirsch, 2002). This credibility mechanism is supported by a number of research studies. For example, people in one study saw video clips of themselves that apparently were filmed during an earlier phase of the study (Nash, Wade, & Brewer, 2009). In each of these clips, the participant was seen passively observing another person who performed several simple actions. In a subsequent memory test, participants often reported that they clearly remembered observing and performing those actions in the initial phase of the study, but in truth they did not; the video clips had been doctored (see also Nash, Wade, & Lindsay, 2009). Seeing themselves in the video lent credibility to the suggestion that they had been in the room when the actions were performed, and thus increased their suggestibility. In a related study, people saw a doctored photo depicting apparent unrest at a protest that genuinely took place but was in reality peaceful (Sacchi et al., 2007). Compared with people who saw the genuine photo, which depicted no unrest, those who saw the doctored photo subsequently recalled the protest having involved more protesters, more arrests, more injuries, and even deaths. The authors proposed that the credibility of the photo was a vital ingredient that influenced people's beliefs and memories.

But credibility is not the whole picture. In a study by Wade et al. (2002), many young adults developed false beliefs and memories of a fictional hot-air balloon ride from their childhood as a result of seeing doctored photos that depicted their childhood selves in a hot-air balloon alongside a family member. Whereas credibility no doubt played a role, the authors suggested that the photos may also have served as "springboards" by helping people to generate perceptually rich information in their minds about what the balloon ride might have been like. The vivid visual detail contained in photographs may support our imaginations and thus may make it easier for us to confuse fantasy with reality—an explanation mentioned earlier with reference to Lindsay et al.'s (2004) study involving genuine photos and false memories of a slime prank. One clear counterpoint to this argument, though, is that photos can also constrain our mental imagery, limiting us to thinking of past events in the way they are depicted pictorially instead of allowing our imaginations to run wild (Garry & Wade, 2005; Hessen-Kayfitz & Scoboria, 2012; Wade et al., 2010).

Even in Nash, Wade, and Brewer's (2009) study, the effect of seeing doctored videos was not necessarily solely due to the credibility of those videos; the data suggested that the effect likely stemmed in part from people's familiarity with having seen certain actions being depicted before, which made them later confuse themselves into thinking they themselves had performed those actions. And in a recent series of studies, Nash (2018) replicated the finding that doctored photos can shape people's beliefs about past public events but found that these distortions occur even when people know that the photos are fake. This finding suggests that the familiarity that arises as a result of viewing photos, even when their credibility is low, contributes to false beliefs and memories. Also consider that people have limited ability to even detect the difference between original and doctored photos (Nightingale et al., 2017), making their potential impact even more alarming.

Further support for the role of familiarity is borne out in a number of other recent studies that look not at how viewing photos influences memory but at how it influences the way we process information and make judgments. These studies show that photos can deliver a subjective feeling of "truth" to the ideas or suggestions that they accompany, even when the photos provide no probative evidence whatsoever (Cardwell et al., 2016; Fenn et al., 2013; Newman et al., 2018). For example, when a trivia statement was presented alongside a related yet uninformative photo, people were more likely to rate the statement as true (Newman et al., 2015). When people were shown a series of celebrity names and asked to agree or disagree that either "This famous person is alive" or "This famous person is dead," they were more likely to confirm the statements as true—regardless of which statement they saw—if a photo of the celebrity was also shown. Again, this was the case even though the photos provided no information about the celebrities' living status (Newman et al., 2012). In similar studies, viewing photos even made people more likely to believe that specific events will occur in the future (Newman et al., 2018). These studies tell us that one further reason why photos might distort our memories is that they can lead us to accept ideas and suggestions as being real. This may well be because photos are so easy for us to mentally process, and, consequently, they deliver a hazy yet misleading feeling of familiarity (Newman et al., 2015).

HOW DOES THE ACT OF TAKING PHOTOS AFFECT ATTENTION AND MEMORY?

Whereas seeing and reviewing photos clearly has both positive and negative consequences for memory, it is also important to look at how the act of actually taking a photo can influence the way we attend to and encode events in memory. On the one hand, taking photos might help people remember more of their experiences because choosing what to photograph, lining up the shot, and physically taking the photo may focus attention. Photo-taking might help us allocate our attention to the scene and to our experience in part because photographing something is a more active process than just observing it (see, e.g., Roediger & Zaromb, 2010). Support for this notion comes from a study in which children and adults took photos with digital cameras while touring a museum (Pathman et al., 2011). They also viewed a slideshow of photos taken by other people for additional exhibits; thus, each participant engaged in both a photo-taking and a photo-viewing task. On a memory test 1 to 2 days later, people recognized more of the photos they themselves had taken on the museum tour than of the photos they merely viewed in the slideshow.

On the other hand, taking photos could harm memory, and this might occur for several reasons. Photographing objects and scenes forces people to divide their attention between the camera and what is right in front of them. This divided attention can lead to less effective encoding similar to what occurs when people use cell phones or laptops while engaged in other activities (Fried, 2008; Hembrooke & Gay, 2003; Hyman et al., 2010; T. S. Smith et al., 2011). Another way that photo-taking could be detrimental to memory is that people may pay less attention to a scene and process it less extensively and meaningfully because they treat the camera as an external storage device, thus, in essence, "outsourcing" their memory by mentally counting on the camera to "remember" for them. This notion is consistent with certain prior work, which shows that people remember less information when they expect to have future access to the information via the internet or because it is saved on an external storage device, such as a computer (Sparrow et al., 2011). Cognitive offloading of this sort appears to encourage people to treat the internet as an extension of their own memory and cognitive systems (Finley et al., 2018; Fisher et al., 2015; Risko & Gilbert, 2016; Ward, 2013), and there are both benefits and costs to doing so. Accordingly, over the past several years, a small but growing body of research has shed light on how and when taking photos can harm memory and how and when they can benefit memory. This research is reviewed in the following sections.

The Photo-Taking Impairment Effect on Memory

One study showing the negative effects of taking photos took its inspiration from the common experience in which people move from one work of art to another in a museum, as if their goal were to capture photos as trophies rather than to appreciate the art itself. In that study, people took a guided tour of a museum and viewed various pieces of art, including sculptures, paintings, pottery, and jewelry (Henkel, 2014). After viewing each object for a fixed amount of time, they were directed by the tour guide to take photos of some of the artwork and to merely look at—but not photograph—other artwork. Participants were not allowed to review the photos after taking them, a situation that emulates the common experience of people being too overwhelmed by the number of photos they take to actually review them (Bowen & Petrelli, 2011; Nunes et al., 2009; Shutterfly, 2014).

A day later, people's memory for the tour was assessed, and the results showed a *photo-taking impairment effect*. People recognized fewer of the works of art and fewer visual details about them when they had photographed them rather than just looked at them. In addition, their memory for which room a given artwork was located in was also impaired when they took photos rather than just looked at those objects. These results suggest that under some circumstances, people might treat the camera as an external storage device, expecting the camera to capture the visual information for them and therefore not attending to it or processing it in a way that would enhance its memorability.

The relationship between photo-taking and memory is complex, though, and can be affected by many factors. Several studies have replicated the photo-taking impairment effect but have found that it more likely is attributable to attentional demands and attentional disengagement involved in taking the photos rather than due exclusively to offloading memory (Niforatos et al., 2017; Soares & Storm, 2018). Studies have found that people more accurately remember which photos were scenes they had seen earlier when they took photos using an ephemeral photo app that they believed did not store the photos, presumably causing them to pay more attention to the scenes because they were not counting on the camera to be an external memory source (van Nimwegen & Bergman, 2019; see also Macias et al., 2015). These findings highlight the interplay between attention and memory.

However, in a series of five experiments—one conducted in an actual museum and four, in a computer simulation of a museum—two of the studies replicated the photo-taking impairment but three did not, and collectively the effect size of the memory impairment was estimated to be relatively small (Nightingale et al., 2017). Differences in subject populations, their motivation and expectations, and the relative levels of attention and memory incurred might contribute to the mixed findings, and more research is needed to better understand how and when taking photos negatively impacts memory.

In addition, the way in which one photographs objects and scenes plays an important role in whether taking photos hurts memory or not. In the second study in Henkel's (2014) museum study, people took photos of some works of art in their entirety, took partial photos of other artwork by zooming in on a specified part (e.g., the feet of the statue, the basket on the street in a painting), and just looked at others. The results replicated the photo-taking impairment effect for artwork photographed as a whole; that is, people remembered fewer of the works of art they photographed and fewer details about those works than for artwork they just looked at. However, the pattern of findings was different for artwork that people had zoomed in on. Specifically, their accuracy in remembering the zoomed-in-on artwork and the visual details of those works of art was just as high as when they merely looked at them. Perhaps because people did not expect the camera to "remember" for them in these circumstances, taking a zoomed-in-on photo did not impair their memory. In addition, it was also found that people's memory for visual details about the artwork was just as accurate for the features they had zoomed in on as for features they did not zoom in on, thus highlighting a key difference between cameras and human information processing. That is, whereas a camera is only capable of recording what it is aimed at, people can create a representation for a whole object while physically directing their visual attention to only one part of it. This point is reminiscent of research on *boundary extension*, an effect that occurs when people observe a conscribed portion of a visual scene, but afterward, they nevertheless activate and remember the larger context of the scene (Hubbard et al., 2010; Intraub et al., 1992). Hence, the type of attention and the mental representations that people create while photographing objects in different ways and with different expectations can impact whether taking photos hurts memory or not.

Consistent with this argument, research has shown that when photo-taking puts demands on attentional and cognitive resources (e.g., when taking a large quantities of photos, when using relatively difficult cameras requiring additional visual attention and planning, when making decisions about actively saving or deleting certain shots), enjoyment and engagement decline (Diehl et al., 2016), thus negatively impacting people's memory for the experiences (Barasch et al., 2018; Nardini et al., 2019; Tamir et al., 2018). In addition, taking photos can negatively impact memory in other ways: When people took photos during a tour, they remembered less of the factual information heard on the tour than when they did not take photos (Zauberman et al., 2015).

The Positive Impact on Memory of Choosing What to Photograph

The choice of whether and what to photograph appears to be yet another important factor in determining whether photo-taking affects subsequent memory for the experience. Whereas people in earlier studies showed a photo-taking impairment effect after being told which artworks to photograph and which to merely look it (Henkel, 2014), more recent research has shown that the freedom to choose what to photograph can benefit memory (Barasch et al., 2017). The act of choosing what to capture in a photo, otherwise known as *volitional photo-taking*, is valuable to study because, in everyday life, people are selective in whether they want to take photos at all and what they want to photograph. In this research, people took a self-guided museum tour with an audio guide that provided factual information about the exhibits, and they either took photos of as many objects as they chose to or were instructed not to take any photos. Participants were later given a recognition test that assessed their memory for visual details about the objects and for auditory information presented in the narrated audio guide. With volitional photography, photo-taking benefited later memory: People who freely took photos while on the tour recognized more visual information than did people who took no photos. These findings were replicated in computer-based laboratory simulations of visiting an art gallery and taking a sightseeing tour.

Volitional photography has costs, though. Taking photos led to reduced retention of auditory information from the narrated guide, thus showing that even volitional photo-taking can distract attention away from some central aspects of one's experience (Barasch et al., 2017). Volitional photo-taking directs attention to those aspects that the viewers deem photo worthy, "in essence rendering visual content primary" (Barasch et al., 2017, p. 1065), but at the same time, it can divert attention away from important, nonvisual aspects.

Furthermore, across their experiments, Barasch et al. (2017) showed that people who took photos volitionally subsequently remembered more visual information for objects they chose to photograph than for objects they chose not to photograph. Photo-takers' memory for nonphotographed objects was somewhat better than the memory of people who took no photos at all, but this pattern was not statistically significant in several of the studies. Taken together, the findings highlight the important role that attention plays in shaping memory, and they provide some support for the idea that taking photos during an activity promotes heightened attention overall and engages more cognitive processes than does having the experience while not taking photos.

The earlier research on nonvolitional photography showed that instructing people what to photograph on a museum tour impaired memory for the objects they encountered and for visual details of those objects (Henkel, 2014). This finding was taken as support for the idea that people offload their memories onto their cameras, expecting their cameras to remember for them (see also Sparrow et al., 2011; Storm & Stone, 2015). Because the photo-taking impairment effect was not found when people freely chose what to take photos of, to better understand the mechanisms, one study manipulated people's expectations regarding the later accessibility of their photos (Study 3; Barasch et al., 2017). Thus, some of those who took photos were told that their photos would be saved on the camera, and others were told their photos would be deleted. As a control, some did not take photos at all. The results showed that regardless of whether people were told their photos would be saved or deleted, people who took photos remembered more visual information than did people who did not take photos.

These results conflict with the idea that people offload their memories to the camera, although there are many reasons why that might be the case here but not in the earlier work. It might be, for instance, that people paid no attention to the instructions that informed them about their photos being saved or deleted, or perhaps they simply did not believe that information. It also might be that when people are engaged in an experience and have heightened enjoyment—which prior work shows occurs when people freely choose what to photograph (Diehl et al., 2016)—they are less likely to off-load their memory. In addition, although people did freely choose what to photograph, they had no control over whether they were assigned to be in the photo-taking condition or the control condition. Thus, although people typically freely choose what to take photos of, the heightened attention brought about simply by taking part in an experiment and being invited to take photos may contribute to the effects observed. The roles of motivation and heightened attentional processes are seen in research in which people did not actually take photos at all but simply planned on doing so and imagined doing so (Diehl et al., 2016). So, it may not be that taking photos per se increases attention but that asking people to take photos increases attention to visual details.

Also consider that in the everyday world, there are certainly situations in which people choose to document an experience with photos, but they take the photos mindlessly with little engagement or are distracted by other agendas, such as posting to social media sites. In those situations, it may be the case that volitionally taking photos harms memory. For instance, research has shown that when people take photos with the intention of sharing them with others on social media sites, they actually enjoy the experiences they captured in the photos less because sharing photos increases their concerns about self-presentation and how others judge them (Barasch et al., 2018). Other work has found that when the act of taking photos places heavier demands attentional or cognitive resources—such as when taking a large number of photos, when using cameras that require heightened visual focus and planning, or when making tough decisions on whether to actively save or delete certain shots—it does not foster enjoyment (Study 7; Diehl et al., 2016). More research is needed to better understand the roles that these factors play in shaping memory and in determining whether and when people do indeed offload their memories onto their cameras.

And despite the rallying cry that relying on technology is ruining people's ability to think and remember, it is important to note that people were externalizing their memories long before computers, cameras, and the internet by using handwritten and typed notes, diaries, calendars, audio recordings, alarms, and other physical reminders. Indeed, people were expressing concerns about the psychological side effects of those advances well ahead of the digital age (Madan, 2014; Nestojko et al., 2013). Consider also the potential positive consequences of offloading memory onto technology, as examined by Storm and Stone (2015). Noting that saving information on computers or cameras might make it more difficult to remember that information unaided (Henkel, 2014; Sparrow et al., 2011), the authors asked whether it might simultaneously benefit people's ability to encode and remember other information. Participants studied the contents of a file containing word lists, not knowing whether they would be asked to save the file or not. Results showed that saving one computer file before studying a new file enhanced people's memory for the new file. This result shows that when people delegate some part of their information processing onto external storage devices, it can enable them to use their cognitive resources for other information and tasks. This *saving-enhanced memory* did not occur, though, when the saving process was viewed as unreliable, which suggests that people strategically offload memory onto external storage devices to reallocate their limited and valuable cognitive resources away from the no-longer-needed information and more efficiently focus their attention and cognitive processing on newer or more important information. Perhaps that is also the case with photos: If we expect the camera to "remember" for us when we take a photo, then we may not engage so much in additional processing and might instead focus greater attention on the next scene or event.

SUMMARY AND A SNAPSHOT OF FUTURE DIRECTIONS

This chapter explored attention economics in the context of how taking and viewing photos impacts memory and cognition. People take photos for many different reasons, such as sharing them with others either in person or via social media, or to just document their experiences. But a main motivator for taking photos is people's underlying expectation that they will later review them to refresh and reactivate their memories.

Drawing on a rich body of research in laboratory and field settings examining participants of all ages and both those with and without memory problems, evidence shows that the richness and specificity of photos provide concrete and distinctive retrieval cues that can indeed help people reactivate and remember their experiences. But the resulting accumulation of a staggeringly high number of photos—made possible through modern digital technology—places new demands on the attentional and cognitive resources that are needed to organize and access their photos. These demands therefore likely reduce the intended effectiveness of photos as retrieval cues.

A growing body of research has also examined situations in which the act of taking photos can either benefit or hurt subsequent memory. Some evidence suggests that when people are tasked with taking photos and directed on which shots to take, taking photos impaired people's memory for what they experienced. One explanation for this photo-taking impairment effect is that people treat cameras as external memory devices and offload their memories onto the camera because of limitations in their attentional resources. However, other research suggests that the photo-taking impairment effect might be relatively small or contingent on other variables that are as yet poorly understood. In addition, when people freely choose what to photograph and are engaged with the task, benefits arise from taking photos, including enhanced memory retention for visual aspects of the events. Putting these findings together, research suggests that the benefits and costs of taking photos while experiencing events can play out dynamically with trade-offs in attention to particular aspects of one's experiences. The increase in attention to visual information and its resulting boost in memory come with costs, for instance, reducing attention to and memory for auditory and contextual information. Thus, overall, there are both positive and negative consequences for memory of taking and viewing one's own photos. It would be useful for future research to examine these costs and benefits in populations that might especially need additional cognitive support, such as older adults with declining memory abilities or people with significant cognitive and memory impairments (e.g., Alzheimer's disease).

Given limited attentional and cognitive resources, how might people wisely and economically "spend" these resources? Scientific research in this area is still developing, and as technology continues to evolve at rapid rate, there is certainly no magic bullet to definitively answer that question. But several factors appear to be especially valuable to consider. The research thus far indicates that how we attend to our environment while taking photos is important. People should be aware of this and make choices informed by the pros and cons. For example, when people focus their attention on visual aspects as they enjoy and engage with an experience, taking photos might have a positive impact on memory, but when situations are more demanding, taking photos might impair memory as might taking photos with the sole focus of sharing them with others as "trophies." It is important to recognize that not all of the photos we take are for the purposes of preserving our memories (e.g., van Dijck, 2008), and flexibly adopting our photo-taking behaviors to suit the varied purposes might be helpful. Moreover, whereas we have primarily focused here on static photographs, one might speculate that many of these attentional and memory effects would be more prominent in the case of taking and reviewing video recordings of one's experiences. Relatively few studies are yet available that speak directly to this speculation, and this would be an interesting direction for further study.

Another factor for people to consider as they choose whether and what to photograph is how much to photograph. The accumulation of large quantities of photos is a major issue, given that digital overload prevents people from finding and using the potentially rich memory cues of photos in their intended fashion. Sometimes less really might be more! Learning about options for photo organization may allow people to make more informed decisions that will help them better achieve their purposes in taking photos, at least when the intended purpose of these photos is memory oriented.

Active engagement with the photos themselves—spending time actually looking at them, reflecting, and reminiscing alone or with others—is another critical ingredient in maximizing the memory cueing potential of photos, and numerous studies show an added benefit: that reminiscing can deliver improvements in well-being and mental health (Westerhof et al., 2010). A critical determinant of subsequent memorability is not just looking at photos but scaffolding memory with the types of cognitive processing that can contribute to stronger, more durable, and more accessible memories. For instance, although learning experiences in museums can be enhanced when children take photos, they need structured exercises and tasks to guide them for the media-augmented learning to be effective (Hillman et al., 2016). Developing the skill to effectively and efficiently document and review past experiences could benefit from technology designed to do just that. The area of "technology-mediated memory" draws from principles about how human memory works to encourage active reflection and has shown promising findings in terms of increases in well-being and memory functioning (Isaacs et al., 2013; Konrad et al., 2016; see also Crete-Nishihata et al., 2012). Such work is still in its infancy, but a future in which evidence-based practices are steeped in theory about how memory works looks promising, especially with growing interest in "lifelogging" in which people use wearable cameras and other devices to document and record sizable portions of their daily experiences (e.g., Gurrin, Smeaton, & Doherty, 2014; Whittaker et al., 2012).

More research is needed to further our understanding of the complex relation between attention and memory when it comes to taking and viewing photos. For example, the content of a photo is likely a significant determinant of its memory-cueing value at a later time, but little is yet known about that. Decades later, a closeup photo of a plate of food or one's toes in front of the ocean might not be sufficiently specific or detailed enough to conjure up rich recollections of the specific meal or trip depicted. The possibility of generational differences is also important to consider. Younger people who are often considered "digital natives"—in that they grew up with computers and the internet in their lives from an early age—may use different information processing styles when reading content and browsing the internet than do people of older generations (Loh & Kanai, 2016). How these groups' attention and memory are influenced by taking and viewing photos may differ in turn (see George & Odgers, 2015, for information on whether mobile technologies may be influencing children's developing brains).

Finally, consider that people often treat photos as if they were a veridical representation of what happened—that the photo *is* reality. But just as perception is a construction of reality, and memory in turn is a reconstruction of perceptual and reflective processes, photos are not necessarily the unfiltered truth. Not only can photos can be altered via cropping, lighting, and filters, they can be digitally manipulated to include elements that were not there at all or to change elements that were. In addition, photos sometimes tell only one small part of the full scene: the part the camera was aimed at. Imagine walking through a museum while wearing a camera that automatically took snapshots as you walked. The camera captures the rich assortment of artwork in its path. But what if you were looking to your side talking to your partner for the better part of the tour? What you attended to, what you thought about, and what you might subsequently remember are not captured by the photos. Similarly, a police officer wearing a body camera might capture some element of a crime that happened, but the officer's attention and gaze might allow them to process and act on much more than what is in the camera's line of sight. In this way, cameras can never rival what the human brain is capable of with its sophisticated perceptual, attentional, and memory processes. And although photos are valuable, they, like memories, are still only interpretations of reality.

REFERENCES

Aschermann, E., Dannenberg, U., & Schulz, A.-P. (1998). Photographs as retrieval cues for children. *Applied Cognitive Psychology*, *12*(1), 55–66. https://doi.org/10.1002/(SICI)1099-0720(199802)12:1<55::AID-ACP490>3.0.CO;2-E

Atchley, P., & Lane, S. (2014). Cognition in the attention economy. In B. H. Ross (Ed.), *Psychology of learning and motivation* (Vol. 61, pp. 133–177). Academic Press. https://doi.org/10.1016/B978-0-12-800283-4.00004-6

Barasch, A., Diehl, K., Silverman, J., & Zauberman, G. (2017). Photographic memory: The effects of volitional photo taking on memory for visual and auditory aspects of an experience. *Psychological Science*, *28*(8), 1056–1066. https://doi.org/10.1177/0956797617694868

Barasch, A., Zauberman, G., & Diehl, K. (2018). How the intention to share can undermine enjoyment: Photo-taking goals and evaluation of experiences. *Journal of Consumer Research*, *44*(6), 1220–1237. https://doi.org/10.1093/jcr/ucx112

Barnard, P. J., Murphy, F. C., Carthery-Goulart, M. T., Ramponi, C., & Clare, L. (2011). Exploring the basis and boundary conditions of SenseCam-facilitated recollection. *Memory*, *19*(7), 758–767. https://doi.org/10.1080/09658211.2010.533180

Blandón-Gitlin, I., & Gerkens, D. (2010). The effects of photographs and event plausibility in creating false beliefs. *Acta Psychologica*, *135*(3), 330–334. https://doi.org/10.1016/j.actpsy.2010.08.008

Bourgeois, M. S., Dijkstra, K., Burgio, L., & Allen-Burge, R. (2001). Memory aids as an augmentative and alternative communication strategy for nursing home residents with dementia. *Augmentative and Alternative Communication*, *17*(3), 196–210. https://doi.org/10.1080/aac.17.3.196.210

Bowen, S., & Petrelli, D. (2011). Remembering today tomorrow: Exploring the human-centered design of digital mementos. *International Journal of Human-Computer Studies*, *69*(5), 324–337. https://doi.org/10.1016/j.ijhcs.2010.12.005

Braun, K. A., Ellis, R., & Loftus, E. F. (2002). Make my memory: How advertising can change our memories of the past. *Psychology & Marketing, 19*(1), 1–23. https://doi.org/10.1002/mar.1000

Brindley, R., Bateman, A., & Gracey, F. (2011). Exploration of use of SenseCam to support autobiographical memory retrieval within a cognitive-behavioural therapeutic intervention following acquired brain injury. *Memory, 19*(7), 745–757. https://doi.org/10.1080/09658211.2010.493893

Brown, A. S., & Marsh, E. J. (2008). Evoking false beliefs about autobiographical experience. *Psychonomic Bulletin & Review, 15,* 186–190. https://doi.org/10.3758/PBR.15.1.186

Browne, G., Berry, E., Kapur, N., Hodges, S., Smyth, G., Watson, P., & Wood, K. (2011). SenseCam improves memory for recent events and quality of life in a patient with memory retrieval difficulties. *Memory, 19*(7), 713–722. https://doi.org/10.1080/09658211.2011.614622

Burt, C. B., Mitchell, D. A., Raggatt, P. F., Jones, C. A., & Cowan, T. M. (1995). A snapshot of autobiographical memory retrieval characteristics. *Applied Cognitive Psychology, 9*(1), 61–74. https://doi.org/10.1002/acp.2350090105

Cabeza, R., Prince, S. E., Daselaar, S. M., Greenberg, D. L., Budde, M., Dolcos, F., LaBar, K. S., & Rubin, D. C. (2004). Brain activity during episodic retrieval of autobiographical and laboratory events: An fMRI study using a novel photo paradigm. *Journal of Cognitive Neuroscience, 16*(9), 1583–1594. https://doi.org/10.1162/0898929042568578

Cardwell, B. A., Henkel, L. A., Garry, M., Newman, E. J., & Foster, J. L. (2016). Nonprobative photos rapidly lead people to believe claims about their own (and other people's) pasts. *Memory & Cognition, 44,* 883–896. https://doi.org/10.3758/s13421-016-0603-1

Ceroni, A. (2018). Personal photo management and preservation. In V. Merzaris, C. Niederee, & R. Logie (Eds.), *Personal multimedia preservation* (pp. 279–314). Springer International. https://doi.org/10.1007/978-3-319-73465-1_8

Chalfen, R. (1998). Family photograph appreciation: Dynamics of medium, interpretation and memory. *Communication & Cognition, 31*(2–3), 161–178.

Chow, T. E., & Rissman, J. (2017). Neurocognitive mechanisms of real-world autobiographical memory retrieval: Insights from studies using wearable camera technology. *Annals of the New York Academy of Sciences, 1396,* 202–221. https://doi.org/10.1111/nyas.13353

Collins, A. M., & Loftus, E. F. (1975). A spreading-activation theory of semantic processing. *Psychological Review, 82*(6), 407–428. https://doi.org/10.1037/0033-295X.82.6.407

Crete-Nishihata, M., Baecker, R. M., Massimi, M., Ptak, D., Campigotto, R., Kaufman, L. D., Brickman, A. M., Turner, G. R., Steinerman, J. R., & Black, S. E. (2012). Reconstructing the past: Personal memory technologies are not just personal and not just for memory. *Human-Computer Interaction, 27*(1–2), 92–123.

Deffenbacher, K. A., Bornstein, B. H., & Penrod, S. D. (2006). Mugshot exposure effects: Retroactive interference, mugshot commitment, source confusion, and unconscious transference. *Law and Human Behavior, 30*(3), 287–307. https://doi.org/10.1007/s10979-006-9008-1

Deocampo, J. A., & Hudson, J. A. (2003). Reinstatement of 2-year-olds' event memory using photographs. *Memory, 11*(1), 13–25. https://doi.org/10.1080/741938177

Diehl, K., Zauberman, G., & Barasch, A. (2016). How taking photos increases enjoyment of experiences. *Journal of Personality and Social Psychology, 111*(2), 119–140. https://doi.org/10.1037/pspa0000055

Fenn, E., Newman, E. J., Pezdek, K., & Garry, M. (2013). The effect of nonprobative photographs on truthiness persists over time. *Acta Psychologica, 144*(1), 207–211. https://doi.org/10.1016/j.actpsy.2013.06.004

Finley, J. R., Brewer, W. F., & Benjamin, A. S. (2011). The effects of end-of-day picture review and a sensor-based picture capture procedure on autobiographical memory using SenseCam. *Memory, 19*(7), 796–807. https://doi.org/10.1080/09658211.2010.532807

Finley, J. R., Naaz, F., & Goh, F. W. (2018). *Memory and technology: How we use information in the brain and the world.* Springer International. https://doi.org/10.1007/978-3-319-99169-6

Fisher, M., Goddu, M. K., & Keil, F. C. (2015). Searching for explanations: How the internet inflates estimates of internal knowledge. *Journal of Experimental Psychology: General, 144*(3), 674–687. https://doi.org/10.1037/xge0000070

Fivush, R., Hudson, J., & Nelson, K. (1984). Children's long-term memory for a novel event: An exploratory study. *Merrill-Palmer Quarterly, 30*(3), 303–316.

Fried, C. B. (2008). In-class laptop use and its effects on student learning. *Computers & Education, 50*(3), 906–914. https://doi.org/10.1016/j.compedu.2006.09.006

Garry, M., Strange, D., Bernstein, D. M., & Kinzett, T. (2007). Photographs can distort memory for the news. *Applied Cognitive Psychology, 21*(8), 995–1004. https://doi.org/10.1002/acp.1362

Garry, M., & Wade, K. A. (2005). Actually, a picture is worth less than 45 words: Narratives produce more false memories than photographs do. *Psychonomic Bulletin & Review, 12*, 359–366. https://doi.org/10.3758/BF03196385

George, M. J., & Odgers, C. L. (2015). Seven fears and the science of how mobile technologies may be influencing adolescents in the digital age. *Perspectives on Psychological Science, 10*(6), 832–851. https://doi.org/10.1177/1745691615596788

Good, J. (2011, September 15). *How many photos have ever been taken?* 1000Memories. http://www.Blog.1000memories.com

Gurrin, C., Smeaton, A. F., & Doherty, A. R. (2014). Lifelogging: Personal big data. *Foundations and Trends in Information Retrieval, 8*(1), 1–125. https://doi.org/10.1561/1500000033

Harrison, B. (2002). Photographic visions and narrative inquiry. *Narrative Inquiry, 12*(1), 87–111. https://doi.org/10.1075/ni.12.1.14har

Hartman, H. (2014, November). *The photo management challenge* [White paper]. Available for download on https://www.suite48a.com/management

Hembrooke, H., & Gay, G. (2003). The laptop and the lecture: The effects of multitasking in learning environments. *Journal of Computing in Higher Education, 15*, 46–64. https://doi.org/10.1007/BF02940852

Henkel, L. A. (2011). Photograph-induced memory errors: When photos make people claim they've done things they haven't. *Applied Cognitive Psychology, 25*(1), 78–86. https://doi.org/10.1002/acp.1644

Henkel, L. A. (2014). Point-and-shoot memories: The influence of taking photos on memory for a museum tour. *Psychological Science, 25*(2), 396–402. https://doi.org/10.1177/0956797613504438

Henkel, L. A., & Carbuto, M. (2008). How source misattributions arise from verbalization, mental imagery, and pictures. In M. Kelley (Ed.), *Applied memory* (pp. 213–234). Nova Science Publishers.

Hessen-Kayfitz, J., Scoboria, A., & Nespoli, K. (2017). The labeling of photos when suggesting false childhood events can enhance or suppress false memory formation. *Psychology of Consciousness: Theory, Research, and Practice, 4*(3), 288–297. https://doi.org/10.1037/cns0000100

Hessen-Kayfitz, J. K., & Scoboria, A. (2012). False memory is in the details: Photographic details differentially predict memory formation. *Applied Cognitive Psychology, 26*(3), 333–341. https://doi.org/10.1002/acp.1839

Heyden, T. (2014, October 14). The old photos helping trigger memories in people with dementia. *BBC News Magazine.* https://www.bbc.com/news/magazine-29596805

Heyman, S. (2015, July 29). Photos, photos, everywhere. *The New York Times.* Retrieved from https://www.nytimes.com/2015/07/23/arts/international/photos-photos-everywhere.html

Hillman, T., Weilenmann, A., Jungselius, B., & Lindell, T. L. (2016). Traces of engagement: Narrative-making practices with smartphones on a museum field trip. *Learning, Media and Technology, 41*(2), 351–370. https://doi.org/10.1080/17439884.2015.1064443

Hodges, S., Berry, E., & Wood, K. (2011). SenseCam: A wearable camera that stimulates and rehabilitates autobiographical memory. *Memory, 19*(7), 685–696. https://doi.org/10.1080/09658211.2011.605591

Hubbard, T. L., Hutchison, J. L., & Courtney, J. R. (2010). Boundary extension: Findings and theories. *Quarterly Journal of Experimental Psychology, 63*(8), 1467–1494. https://doi.org/10.1080/17470210903511236

Hudson, J. A., & Fivush, R. (1991). As time goes by: Sixth graders remember a kindergarten experience. *Applied Cognitive Psychology, 5*(4), 347–360. https://doi.org/10.1002/acp.2350050405

Hyman, I., Jr., Boss, S., Wise, B. M., McKenzie, K. E., & Caggiano, J. M. (2010). Did you see the unicycling clown? Inattentional blindness while walking and talking on a cell phone. *Applied Cognitive Psychology, 24*(5), 597–607. https://doi.org/10.1002/acp.1638

Ianì, F., Mazzoni, G., & Bucciarelli, M. (2018). The role of kinematic mental simulation in creating false memories. *Journal of Cognitive Psychology, 30*(3), 292–306. https://doi.org/10.1080/20445911.2018.1426588

Instagram company statistics. (2016, September 1). Statistic Brain Research Institute. http://www.statisticbrain.com/instagram-company-statistics/

Intraub, H., Bender, R. S., & Mangels, J. A. (1992). Looking at pictures but remembering scenes. *Journal of Experimental Psychology: Learning, Memory, and Cognition, 18*(1), 180–191. https://doi.org/10.1037/0278-7393.18.1.180

Isaacs, E., Konrad, A., Walendowski, A., Lennig, T., Hollis, V., & Whittaker, S. (2013, April). Echoes from the past: How technology mediated reflection improves well-being. In W. E. Mackay (General Chair), *CHI '13: Proceedings of the SIGCHI Conference on Human Factors in Computing Systems* (pp. 1071–1080). Association for Computing Machinery.

Jack, F., Martyn, E., & Zajac, R. (2015). Getting the picture: Effects of sketch plans and photographs on children's, adolescents' and adults' eyewitness recall. *Applied Cognitive Psychology, 29*(5), 723–734. https://doi.org/10.1002/acp.3156

Johnson, M. K., Raye, C. L., Mitchell, K. J., & Ankudowich, E. (2011). The cognitive neuroscience of true and false memories. In R. F. Belli (Ed.), *True and false recovered memories: Toward a reconciliation of the debate: Vol. 58. Nebraska Symposium on Motivation* (pp. 15–52). Springer. https://doi.org/10.1007/978-1-4614-1195-6_2

Konrad, A., Isaacs, E., & Whittaker, S. (2016). Technology mediated memory: Is technology altering our memories and interfering with well-being? *ACM Transactions on Computer–Human Interaction, 23*(4), Article No. 23. https://doi.org/10.1145/2934667

Koutstaal, W., Schacter, D. L., Johnson, M. K., Angell, K. E., & Gross, M. S. (1998). Post-event review in older and younger adults: Improving memory accessibility of complex everyday events. *Psychology and Aging, 13*(2), 277–296. https://doi.org/10.1037/0882-7974.13.2.277

Koutstaal, W., Schacter, D. L., Johnson, M. K., & Galluccio, L. (1999). Facilitation and impairment of event memory produced by photograph review. *Memory & Cognition, 27*, 478–493. https://doi.org/10.3758/BF03211542

Lindner, I., Echterhoff, G., Davidson, P. S. R., & Brand, M. (2010). Observation inflation. Your actions become mine. *Psychological Science, 21*(9), 1291–1299. https://doi.org/10.1177/0956797610379860

Lindsay, D. S., Hagen, L., Read, J. D., Wade, K. A., & Garry, M. (2004). True photographs and false memories. *Psychological Science, 15*(3), 149–154. https://doi.org/10.1111/j.0956-7976.2004.01503002.x

Loh, K. K., & Kanai, R. (2016). How has the internet reshaped human cognition? *Neuroscientist, 22*(5), 506–520. https://doi.org/10.1177/1073858415595005

Loveday, C., & Conway, M. A. (2011). Using SenseCam with an amnesic patient: Accessing inaccessible everyday memories. *Memory, 19*(7), 697–704. https://doi.org/10.1080/09658211.2011.610803

Macias, C., Yung, A., Hemmer, P., & Kidd, C. (2015). Memory strategically encodes externally unavailable information. In D. C. Noelle, R. Dale, A. S. Warlaumont, J. Yoshimi, T. Matlock, C. D. Jennings, & P. P. Maglio (Eds.), *Proceedings of the 37th Annual Meeting of the Cognitive Science Society* (pp. 1458–1463). Cognitive Science Society. https://cogsci.mindmodeling.org/2015/papers/0255/paper0255.pdf

Madan, C. R. (2014). Augmented memory: A survey of the approaches to remembering more. *Frontiers in Systems Neuroscience, 8*, 30. https://doi.org/10.3389/fnsys.2014.00030

Madigan, S. (1974). Representational storage in picture memory. *Bulletin of the Psychonomic Society, 4*, 567–568. https://doi.org/10.3758/BF03334293

Mair, A., Poirier, M., & Conway, M. A. (2017). Supporting older and younger adults' memory for recent everyday events: A prospective sampling study using SenseCam. *Consciousness and Cognition, 49*, 190–202. https://doi.org/10.1016/j.concog.2017.02.008

Mazzoni, G., & Kirsch, I. (2002). Autobiographical memories and beliefs: A preliminary metacognitive model. In T. J. Perfect & B. L. Schwartz (Eds.), *Applied metacognition* (pp. 121–145). Cambridge University Press. https://doi.org/10.1017/CBO9780511489976.007

Meeker, M. (2014, May 28). *Internet trends 2014—Code conference.* Virtual Properties. https://www.virtualproperties.com/blog/g/14/mm/Internet_Trends_2014.pdf

Meeker, M. (2016, June 1). *Internet trends 2016—Code conference.* https://www.cs.rutgers.edu/~badri/552dir/papers/intro/Meeker-2016.pdf

Morgan, C. A., III, Southwick, S., Steffian, G., Hazlett, G. A., & Loftus, E. F. (2013). Misinformation can influence memory for recently experienced, highly stressful events. *International Journal of Law and Psychiatry, 36*(1), 11–17. https://doi.org/10.1016/j.ijlp.2012.11.002

Murphy, F. C., Barnard, P. J., Terry, K. A., Carthery-Goulart, M. T., & Holmes, E. A. (2011). SenseCam, imagery and bias in memory for wellbeing. *Memory, 19*(7), 768–777. https://doi.org/10.1080/09658211.2010.551130

Nardini, G., Lutz, R., & LeBoeuf, R. (2019). How and when taking pictures undermines the enjoyment of experiences. *Psychology & Marketing, 36*(5), 520–529. https://doi.org/10.1002/mar.21194

Nash, R. A. (2018). Changing beliefs about past public events with believable and unbelievable doctored photographs. *Memory, 26*(4), 439–450. https://doi.org/10.1080/09658211.2017.1364393

Nash, R. A., Wade, K. A., & Brewer, R. J. (2009). Why do doctored images distort memory? *Consciousness and Cognition, 18*(3), 773–780. https://doi.org/10.1016/j.concog.2009.04.011

Nash, R. A., Wade, K. A., Garry, M., & Adelman, J. S. (2017). A robust preference for cheap-and-easy strategies over reliable strategies when verifying personal memories. *Memory, 25*(7), 890–899. https://doi.org/10.1080/09658211.2016.1214280

Nash, R. A., Wade, K. A., & Lindsay, D. S. (2009). Digitally manipulating memory: Effects of doctored videos and imagination in distorting beliefs and memories. *Memory & Cognition, 37*, 414–424. https://doi.org/10.3758/MC.37.4.414

Nelson, D. L., Reed, V. S., & Walling, J. R. (1976). Pictorial superiority effect. *Journal of Experimental Psychology: Human Learning and Memory, 2*(5), 523–528. https://doi.org/10.1037/0278-7393.2.5.523

Nestojko, J. F., Finley, J. R., & Roediger, H. L., III. (2013). Extending cognition to external agents. *Psychological Inquiry, 24*(4), 321–325. https://doi.org/10.1080/1047840X.2013.844056

Newman, E. J., Azad, T., Lindsay, D. S., & Garry, M. (2018). Evidence that photos promote rosiness for claims about the future. *Memory & Cognition, 46*, 1223–1233. https://doi.org/10.3758/s13421-016-0652-5

Newman, E. J., Garry, M., Bernstein, D. M., Kantner, J., & Lindsay, D. S. (2012). Nonprobative photographs (or words) inflate truthiness. *Psychonomic Bulletin & Review, 19*, 969–974. https://doi.org/10.3758/s13423-012-0292-0

Newman, E. J., Garry, M., Unkelbach, C., Bernstein, D. M., Lindsay, D. S., & Nash, R. A. (2015). Truthiness and falsiness of trivia claims depend on judgmental contexts. *Journal of Experimental Psychology: Learning, Memory, and Cognition, 41*(5), 1337–1348. https://doi.org/10.1037/xlm0000099

Niforatos, E., Cinel, C., Mack, C., Langheinrich, M., & Ward, G. (2017, June). Can less be more? Contrasting limited, unlimited, and automatic picture capture for augmenting memory recall. *Proceedings of the ACM on Interactive, Mobile, Wearable and Ubiquitous Technologies, 1*(2), Article No. 21. https://doi.org/10.1145/3090086

Nightingale, S. J., Wade, K. A., Watson, D. G., Mills, A. J., Zajac, R., Garry, M., & Henkel, L. A. (2017). *Taking photos probably has a trivial influence on memory for a museum tour* [Unpublished manuscript]. Department of Psychology, University of Warwick.

Noyes, D. (2020, January). The top 20 valuable Facebook statistics—Updated January 2020. https://zephoria.com/top-15-valuable-facebook-statistics/

Nunes, M., Greenberg, S., & Neustaedter, C. (2009). Using physical memorabilia as opportunities to move into collocated digital photo-sharing. *International Journal of Human-Computer Studies, 67*(12), 1087–1111. https://doi.org/10.1016/j.ijhcs.2009.09.007

Paivio, A., Rogers, T. B., & Smythe, P. C. (1968). Why are pictures easier to recall than words? *Psychonomic Science, 11*, 137–138. https://doi.org/10.3758/BF03331011

Parisi, L., & Henkel, L. A. (2014). *Here's looking at me: The effects of seeing oneself in a photo on memory reconstruction* [Unpublished manuscript]. Department of Psychology, Fairfield University.

Pathman, T., Doydum, A., & Bauer, P. J. (2013). Bringing order to life events: Memory for the temporal order of autobiographical events over an extended period in school-aged children and adults. *Journal of Experimental Child Psychology, 115*(2), 309–325. https://doi.org/10.1016/j.jecp.2013.01.011

Pathman, T., Samson, Z., Dugas, K., Cabeza, R., & Bauer, P. J. (2011). A "snapshot" of declarative memory: Differing developmental trajectories in episodic and autobiographical memory. *Memory, 19*(8), 825–835. https://doi.org/10.1080/09658211.2011.613839

Pauly-Takacs, K., Moulin, C. J. A., & Estlin, E. (2011). SenseCam as a rehabilitation tool in a child with anterograde amnesia. *Memory, 19*(7), 705–712. https://doi.org/10.1080/09658211.2010.494046

Risko, E. F., & Gilbert, S. J. (2016). Cognitive offloading. *Trends in Cognitive Sciences, 20*(9), 676–688. https://doi.org/10.1016/j.tics.2016.07.002

Roediger, H. L., & Zaromb, F. M. (2010). Memory for action. In L. Backman & L. Nyberg (Eds.), *Memory, aging and the brain* (pp. 24–52). Psychology Press.

Sacchi, D. L. M., Agnoli, F., & Loftus, E. F. (2007). Changing history: Doctored photographs affect memory for past public events. *Applied Cognitive Psychology, 21*(8), 1005–1022. https://doi.org/10.1002/acp.1394

Salmon, K. (2001). Remembering and reporting by children: The influence of cues and props. *Clinical Psychology Review, 21*(2), 267–300. https://doi.org/10.1016/S0272-7358(99)00048-3

Schacter, D. L., Koutstaal, W., Johnson, M. K., Gross, M. S., & Angell, K. E. (1997). False recollection induced by photographs: A comparison of older and younger adults. *Psychology and Aging, 12*(2), 203–215. https://doi.org/10.1037/0882-7974.12.2.203

Seamon, J. G., Moskowitz, T. N., Swan, A. E., Zhong, B., Golembeski, A., Liong, C., Narzikul, A. C., & Sosan, O. A. (2014). SenseCam reminiscence and action recall in memory-unimpaired people. *Memory, 22*(7), 861–866. https://doi.org/10.1080/09658211.2013.839711

Shepard, R. N. (1967). Recognition memory for words, sentences, and pictures. *Journal of Verbal Learning and Verbal Behavior, 6*(1), 156–163. https://doi.org/10.1016/S0022-5371(67)80067-7

Shutterfly. (2014, October). *Photos and storytelling: Report on survey conducted by Edelman Berland.*

Silva, A. R., Pinho, M. S., Macedo, L., & Moulin, C. J. A. (2018). A critical review of the effects of wearable cameras on memory. *Neuropsychological Rehabilitation, 28*(1), 117–141. https://doi.org/10.1080/09602011.2015.1128450

Silva, A. R., Pinho, S., Macedo, L. M., & Moulin, C. J. (2013). Benefits of SenseCam review on neuropsychological test performance. *American Journal of Preventive Medicine, 44*(3), 302–307. https://doi.org/10.1016/j.amepre.2012.11.005

Smith, C. (2020, February 6). *145 Snapchat statistics, facts and figures (2020): By the numbers.* DMR Business Statistics. Retrieved March 20, 2020, from https://expandedramblings.com/index.php/snapchat-statistics/

Smith, T. S., Isaak, M. I., Senette, C. G., & Abadie, B. G. (2011). Effects of cell-phone and text-message distractions on true and false recognition. *Cyberpsychology, Behavior, and Social Networking, 14*(6), 351–358. https://doi.org/10.1089/cyber.2010.0129

Soares, J. S., & Storm, B. C. (2018). Forget in a flash: A further investigation of the photo-taking-impairment effect. *Journal of Applied Research in Memory and Cognition, 7*(1), 154–160. https://doi.org/10.1016/j.jarmac.2017.10.004

Sparrow, B., Liu, J., & Wegner, D. M. (2011). Google effects on memory: Cognitive consequences of having information at our fingertips. *Science, 333*(6043), 776–778. https://doi.org/10.1126/science.1207745

St. Jacques, P. L., Conway, M. A., & Cabeza, R. (2011). Gender differences in autobiographical memory for everyday events: Retrieval elicited by SenseCam images versus verbal cues. *Memory, 19*(7), 723–732. https://doi.org/10.1080/09658211.2010.516266

St. Jacques, P. L., Montgomery, D., & Schacter, D. L. (2015). Modifying memory for a museum tour in older adults: Reactivation-related updating that enhances and distorts memory is reduced in ageing. *Memory, 23*(6), 876–887. https://doi.org/10.1080/09658211.2014.933241

St. Jacques, P. L., Olm, C., & Schacter, D. L. (2013). Neural mechanisms of reactivation-induced updating that enhance and distort memory. *Proceedings of the National Academy of Sciences, 110*(49), 19671–19678. https://doi.org/10.1073/pnas.1319630110

St. Jacques, P. L., & Schacter, D. L. (2013). Modifying memory: Selectively enhancing and updating personal memories for a museum tour by reactivating them. *Psychological Science, 24*(4), 537–543. https://doi.org/10.1177/0956797612457377

Storm, B. C., & Stone, S. M. (2015). Saving-enhanced memory: The benefits of saving on the learning and remembering of new information. *Psychological Science, 26*(2), 182–188. https://doi.org/10.1177/0956797614559285

Subramaniam, P., Woods, B., & Whitaker, C. (2014). Life review and life story books for people with mild to moderate dementia: A randomised controlled trial. *Aging & Mental Health, 18*(3), 363–375. https://doi.org/10.1080/13607863.2013.837144

Tamir, D., Templeton, E., Ward, A., & Zaki, J. (2018). Media usage diminishes memory for experiences. *Journal of Experimental Social Psychology, 76*, 161–168. https://doi.org/10.1016/j.jesp.2018.01.006

van Dijck, J. F. (2008). Digital photography: Communication, identity, memory. *Visual Communication, 7*(1), 57–76. https://doi.org/10.1177/1470357207084865

van Nimwegen, C., & Bergman, K. (2019). Effects on cognition of the burn after reading principle in ephemeral medial applications. *Behaviour & Information Technology, 38*(10), 1060–1067. https://doi.org/10.1080/0144929X.2019.1659853

Wade, K. A., Garry, M., Nash, R. A., & Harper, D. N. (2010). Anchoring effects in the development of false childhood memories. *Psychonomic Bulletin & Review, 17*, 66–72. https://doi.org/10.3758/PBR.17.1.66

Wade, K. A., Garry, M., Read, J. D., & Lindsay, D. S. (2002). A picture is worth a thousand lies: Using false photographs to create false childhood memories. *Psychonomic Bulletin & Review, 9*, 597–603. https://doi.org/10.3758/BF03196318

Wade, K. A., Nash, R. A., & Garry, M. (2014). People consider reliability and cost when verifying their autobiographical memories. *Acta Psychologica, 146*, 28–34. https://doi.org/10.1016/j.actpsy.2013.12.001

Wade, K. A., Nightingale, S. J., & Colloff, M. F. (2017). Photos and memory. In R. A. Nash & J. Ost (Eds.), *False and distorted memories* (pp. 39–54). Psychology Press.

Ward, A. F. (2013). Supernormal: How the Internet is changing our memories and our minds. *Psychological Inquiry, 24*(4), 341–348. https://doi.org/10.1080/1047840X.2013.850148

Westerhof, G. J., Bohlmeijer, E., & Webster, J. (2010). Reminiscence and mental health: A review of recent progress in theory, research and interventions. *Ageing and Society, 30*(4), 697–721. https://doi.org/10.1017/S0144686X09990328

Whittaker, S., Bergman, O., & Clough, P. (2010). Easy on that trigger dad: A study of long term family photo retrieval. *Personal and Ubiquitous Computing, 14*, 31–43. https://doi.org/10.1007/s00779-009-0218-7

Whittaker, S., Kalnikaite, V., Petrelli, D., Sellen, A., Villar, N., Bergman, O., . . . Brockmeier, J. (2012). Socio-technical lifelogging: Deriving design principles for a future proof digital past. *Human-Computer Interaction, 27*(1–2), 37–62. https://doi.org/10.1080/07370024.2012.656071

Woodberry, E., Browne, G., Hodges, S., Watson, P., Kapur, N., & Woodberry, K. (2015). The use of a wearable camera improves autobiographical memory in patients with Alzheimer's disease. *Memory, 23*(3), 340–349. https://doi.org/10.1080/09658211.2014.886703

Yasuda, K., Kuwabara, K., Kuwahara, N., Abe, S., & Tetsutani, N. (2009). Effectiveness of personalised reminiscence photo videos for individuals with dementia. *Neuropsychological Rehabilitation, 19*(4), 603–619. https://doi.org/10.1080/09602010802586216

Zauberman, G., Silverman, J., Diehl, K., & Barasch, A. (2015). Photographic memory: The effects of the photo-taking on memory for auditory and visual information. In K. Diehl & C. Yoon (Eds.), *Advances in consumer research* (Vol. 43, pp. 218–223). Association for Consumer Research.

6 THE MULTITASKING MOTORIST AND THE ATTENTION ECONOMY

DAVID L. STRAYER, DOUGLAS GETTY, FRANCESCO BIONDI, AND JOEL M. COOPER

Operating a motor vehicle is an everyday activity for millions of adults, and for many, it provides a mode of transportation that is critical to daily living. Driving is also one of the riskiest activities that most adults perform on a regular basis. Motor vehicle crashes are the leading cause of accidental injury deaths in the United States (Liu et al., 2015). Driving provides a unique opportunity to examine the attention economy in an everyday context. Some aspects of driving, such as maintaining lane position on predictable sections of the highway, can become relatively automatic, requiring little attention to be performed well (e.g., Medeiros-Ward et al., 2014). By contrast, reacting to unexpected or unpredictable events requires attention for successful driving performance. The objective of this chapter is to explore the relationship between attention economy and driving. We focus primarily on situations in which the performance of a concurrent nondriving activity adversely affects driving.

Inattentive and distracted driving occurs when a motorist fails to allocate sufficient attention to activities critical to safe driving (Regan et al., 2011; Regan & Strayer, 2014). In many circumstances, this involves diverting attention from driving to a concurrent activity that is unrelated to the safe

https://doi.org/10.1037/0000208-007

Human Capacity in the Attention Economy, S. Lane and P. Atchley (Editors)

operation of the vehicle (e.g., talking or texting on a smartphone). The degree to which driving is altered by a secondary task provides a metric for understanding the relationship between attention and driving (Strayer et al., 2015; Strayer, Cooper, et al., 2017). When a driver attempts to perform an activity unrelated to the primary task of driving, the attention allocated to the driving task decreases. Given the limited resources in the attention economy, there is a reciprocal relationship between the attention allocated to the two tasks: As the processing priority of an unrelated activity increases, the allocation of attention to the driving task decreases (e.g., Kahneman, 1973; Navon & Gopher, 1979).

SPIDER: A FRAMEWORK FOR UNDERSTANDING DISTRACTED DRIVING

Safe driving requires a detailed awareness of the driving environment, which is often information dense and dynamic. Drivers must create and continuously update a mental model of the driving environment that reflects the current driving situation and contains detailed information about the speed and relative position of other vehicles and pedestrians, their own position within a lane, and many other hazards that may present themselves unexpectedly (Cooper et al., 2013; Horrey et al., 2006; Lu et al., 2017).

Situation awareness in driving depends on several mental processes, including *scanning* specific areas for indications of threats, *predicting* potential threats if they are not visible, *identifying* threats and objects in the driving scenario when they occur, *deciding* if an action is necessary, and *executing* appropriate *responses—SPIDER* for short (Fisher & Strayer, 2014; Strayer & Fisher, 2016). SPIDER comprises an active set of psychological processes that depend on the limited capacity of attention (Kahneman, 1973). When drivers engage in secondary activities, attention is diverted from driving, thereby impairing performance of these SPIDER-related processes (Regan et al., 2011; Regan & Strayer, 2014). Figure 6.1 presents the relationship between the allocation of limited capacity attention, the SPIDER-related processes, and their bidirectional links to a driver's situation awareness. SPIDER is important for establishing and maintaining good situation awareness, which is important for coordinating and scheduling the SPIDER-relevant processes.

Visual Scanning

Drivers are responsible for visual scanning of the driving environment, including but not limited to scanning the forward roadway, periphery,

FIGURE 6.1. The SPIDER Model of Attention in Driving

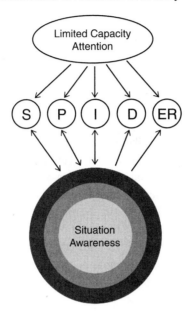

Note. SPIDER stands for scanning, predicting, identifying, deciding, and executing responses. These SPIDER-related processes dependent on limited capacity attention and are critical to a driver's situation awareness. Under distraction, the driver's situation awareness is reduced, as illustrated by progressively smaller and lighter circles. Situation awareness is informed and updated by the SPIDER-related processes (i.e., scanning, predicting, and identifying), and facilitates expectancy-based processing of the driving scene. The loss of situation awareness impairs driving performance and increases the relative risk of a crash.

rear- and side-view mirrors, and vehicle blind spots. Engaging in a secondary task has been shown to decrease visual scanning (Horrey et al., 2006; Victor et al., 2005). Even secondary tasks with no explicit visual component have been shown to increase gaze concentration on the center of the driving scene (Briggs et al., 2017; Engström et al., 2005; Harbluk et al., 2007; He et al., 2011; Recarte & Nunes, 2000; Reimer, 2009; Reimer et al., 2012; Strayer, Cooper, et al., 2017; Tsai et al., 2007; Victor et al., 2005), which might also affect lateral lane position variability (Readinger et al., 2002; Rogers et al., 2005; Wilson et al., 2008; but see Cooper et al., 2013). A change in the typical visual scanning pattern has an obvious detrimental effect on a driver's ability to detect potential hazards in the driving environment (e.g., a pedestrian crossing the roadway).

Hazard Prediction

Anticipatory glances toward a particular area in the roadway is an indicator that a driver is making top-down predictions about possible hazards that may occur at that location. For example, a driver may glance several times at a blind alleyway that they are about to pass by to ensure another vehicle does not drive out in the road in front of them. Several studies have shown that secondary-task engagement reduces anticipatory glances to potential hazards (Biondi et al., 2015; Taylor et al., 2015). An on-road study by Biondi et al. (2015) found that the likelihood of making a glance for pedestrians in a crosswalk was impaired by a cognitively demanding secondary task. Taylor et al. (2015) found that secondary-task engagement made experienced drivers perform similarly to inexperienced drivers. This assertion was recently supported by Wright et al. (2016), who found that experienced drivers look more frequently at the location of potential hazards in the roadway than younger drivers. This finding illustrates how an expectancy-driven search of the driving environment is impaired.

Identification

Drivers engaged in a secondary task may experience *inattentional blindness*, whereby they fail to process items that fall within their field of view (Mack & Rock, 1998; Simons & Chabris, 1999; Strayer & Drews, 2007; Strayer & Johnston, 2001). Strayer et al. (2003) tested drivers' recognition memory for items they looked at while driving (as verified by an eye tracker). They found that drivers who were talking on their cell phone experienced a 50% decrease in recognition memory for items compared with baseline performance. Subsequent studies varied the safety relevance of objects in the driving scene (e.g., pedestrians, other moving vehicles, parked cars, billboards) to see if more safety-relevant objects were protected from inattention blindness; however, there was no relationship between safety relevance and susceptibility to inattention blindness (Strayer et al., 2004). While multitasking, drivers do not appear to strategically prioritize the processing of safety-critical information in the driving scene over the cell phone conversation. Corroborating evidence for inattentional blindness comes from the event-related brain potential (ERP) literature, which has shown that the amplitude of the P300 component of the ERP elicited by driving-relevant events decreases as secondary-task demands increase (e.g., Strayer et al., 2004, 2013; Strayer & Drews, 2007). This ERP pattern shows that secondary-task engagement compromises the initial encoding of traffic-related information.

Decision Making

Good decision making is important for safe driving. For example, drivers must evaluate several sources of information when deciding to change lanes in traffic or make a left turn with oncoming traffic. When drivers divert attention from driving, they often fail to fully evaluate the alternative sources of information. Cooper et al. (2009) observed that when drivers were talking on a hands-free cell phone, they were more likely to make unsafe lane changes. Similarly, Cooper and Zheng (2002) found that drivers misjudged the gap size and speed of oncoming vehicles, and this was most apparent on wet roadways. In both studies, divided attention led to unsafe decision making that increased the risk of a crash.

Execution of a Response

One of the hallmark signatures of divided attention is a slowing of reaction time to imperative events. When something in the driving environment requires a rapid response (e.g., braking in response to a child running across the street), those reactions are often delayed (for an overview of brake reaction time under secondary-task load, see Atchley et al., 2017; Caird et al., 2008; Horrey & Wickens, 2006; Ishigami & Klein, 2009). The effect of secondary-task load is magnified as the perceptual load in the driving environment increases (e.g., when the number of vehicles on the roadway increases, Strayer et al., 2003). The reaction time distribution is not normal but is positively skewed so that the right-hand tail of the distribution is elongated, that is, slow braking responses are particularly slow (Ratcliff & Strayer, 2014). Brown et al. (2001) found that these sluggish brake reactions increase both the likelihood and severity of a motor vehicle collision.

Situation Awareness

In the context of driving, situation awareness reflects a motorist's mental model of the driving environment (e.g., Durso et al., 2007; Endsley, 1995, 2015; Horrey et al., 2006; Kass et al., 2007). The SPIDER-related processes inform and update a driver's situation awareness (i.e., scanning, predicting, and identifying). Performing a secondary task that is unrelated to the primary task of driving competes with the SPIDER-related processes supporting good situation awareness and, consequently, impairs the ability to make good decisions and respond quickly and accurately (Strayer, 2015).

DYNAMIC FLUCTUATIONS IN SITUATION AWARENESS

The attention allocated to the driving task varies over time as a function of the arousal and engagement of the driver, demands of the driving environment, and concurrent performance of any secondary tasks.[1] These fluctuations in attention result in variation in the SPIDER-related processes critical to a driver's situation awareness. Consider the example depicted in Figure 6.2 in which the circumference of the cylinder represents the moment-to-moment level of situational awareness, and a larger circumference indicates higher situational awareness. In the example, a driver diverts attention from the driving task to perform a concurrent secondary task that begins at time (t) t_0 and ends at t_1. Before t_0, full attention is allocated to the driving task (and the SPIDER-related processes), and situation awareness is high. At t_0, the driver begins to perform a secondary task, attention is diverted from the driving environment, situation awareness diminishes, and performance on the driving task is impaired. At t_1, situation awareness reaches its nadir, and the impairments to driving reach their zenith. The curve relating loss of situation awareness and driving impairment is depicted as a continuous function, which has parameters for the rate at which situation awareness is lost and the amount of situation awareness that is lost.[2] These parameters vary as a function of the demands of the primary task (i.e., driving) and secondary task.

At t_1, the secondary task is terminated, and attention is returned to the SPIDER-related processes critical to a driver's situation awareness. However, full situation awareness is not recovered until t_2. The curve relating the recovery of situation awareness and driving impairment is also depicted as a continuous function, which has parameters for the rate at which situation awareness is recovered and the amount of situation awareness that is recovered. The recovery of situation awareness is thought to follow a power function (e.g., Turrill et al., 2016) with greater recovery shortly after the

---

[1] The concentric circles of situation awareness shown in Figure 6.1 can fluctuate dynamically over time. For example, a 90-degree rotation along the north–south axis of the concentric circles of situation awareness shown in Figure 6.1 represents one time slice in the cylinder shown in Figure 6.2. Figure 6.2 shows loss and recovery of situation awareness as symmetrical; however, they are likely governed by different factors, and a symmetrical relationship is only one out of many theoretical possibilities.
[2] For example, if the loss of situation awareness follows a power function, then the rate at which situation awareness is lost over time, $b(x)$, would be scaled by the negative exponent c (e.g., $b(t)^{-c}$, where b is the amount of situation awareness that is lost and c is the rate of loss).

FIGURE 6.2. Dynamic Fluctuations in Situation Awareness (SA) Plotted as a Function of Time (T) When a Motorist Performs an Attention-Demanding Secondary Task

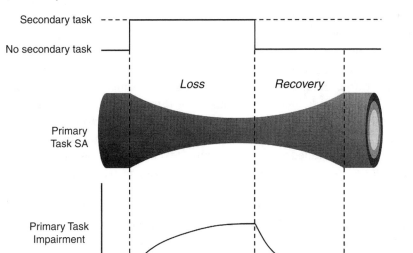

Note. The diameter of the cylinder at any point in time reflects the level of SA; a larger diameter reflects a greater level of SA. A cross-sectional slice of the cylinder at any point in time reflects the SA depicted in the SPIDER (scanning, predicting, identifying, deciding, and executing responses) model (cf. Figure 6.1). In the current figure, SA is high at t_0, the moment that a motorist begins to perform an attention-demanding secondary task. The moment when the secondary task is completed is reflected by t_1. Between t_0 and t_1, SA is systematically lost (reflected by a decreasing diameter of the cylinder), and performance on the primary task of driving progressively degrades. Between t_1 and t_2, SA is slowly recovered (reflected by an increasing diameter of the cylinder).

termination of a secondary task and diminishing returns thereafter. Estimates of the time to recover situation awareness following a secondary task range from 20 to 60 seconds, depending on the complexity of the driving task and the complexity of the secondary task. The result is that effects of a secondary task with duration t_1 has effects on situation awareness and driving performance until t_2. The reduction in situation awareness from t_1 to t_2 reflects a "technology hangover" of sorts.

Let's consider a concrete example of a driver who uses their voice to send a short text message while they are driving. In many current-model-year vehicles, pressing a button on the steering wheel and issuing a voice command to send a text message can accomplish this action. For example, the motorist may wish to text a colleague at work that they are running late. In this example, t_0 reflects the moment the driver pressed the button on the steering wheel. The interval from t_0 to t_1 reflects the time to think about and vocalize the text message, review the message for accuracy, and send it to the desired recipient. The interval from t_1 to t_2 reflects the time to recover any situation awareness lost during the voice-texting interaction. The total period in which situation awareness falls below single-task baseline levels is from t_0 to t_2. Estimates of the t_0 to t_1 interval for simple texts are approximately 30 seconds for voice-based texting (Strayer, Cooper, et al., 2017), and estimates of the residual cost interval (i.e., t_1 to t_2) average approximately 27 seconds (Strayer et al., 2015). Thus, the total time that situation awareness is degraded by sending a simple text message using a voice-based interface is 57 seconds. If a motorist were driving at 62 mph, they would travel a mile during this interaction!

As situation awareness diminishes, motorists' behavior becomes more reactive and less proactive (Braver et al., 2007). Reactive control depends more on bottom-up processing of the driving environment and characterizes decisions made using a limited subset of available information. Low situation awareness leads to poor decision making and slower (or no) reactions by the driver. By contrast, high situation awareness involves more top-down anticipatory processes guided by a fuller sampling of task-relevant information. High situation awareness results in higher quality decisions and faster and more appropriate actions. Consider a scenario in which a driver is confronted with an obstacle in the roadway (e.g., a pothole, roadway debris, a disabled car). On the one hand, a driver with good situation awareness has more options for evasive actions at their disposal (e.g., hit the brakes, swerve left, swerve right) because they are cognizant of the diving conditions, including the vehicles surrounding their car. Alert drivers may also be able to notice other vehicles reacting to the obstacle and take preemptive actions to avoid the hazard altogether. On the other hand, a distracted driver with low situation awareness may only consider one alternative (e.g., hit the brakes), or they may take no evasive action if their situation awareness is sufficiently low that they do not notice the obstacle. Consequently, many of the actions performed by distracted drivers are suboptimal—and lead to postevent "What were they thinking?" questions. They weren't thinking as clearly because lower levels of situation awareness led to a reactive mode of control.

Figure 6.3 provide a metric of driver distraction caused by different levels of secondary-task demand and the subsequent loss and recovery of situation awareness. The left panel (a) illustrates an easy secondary task, and the right panel (b) shows the effects of a mentally demanding secondary

FIGURE 6.3. The Effect of Performing a Mentally Demanding Secondary Task on the Driver's Situation Awareness

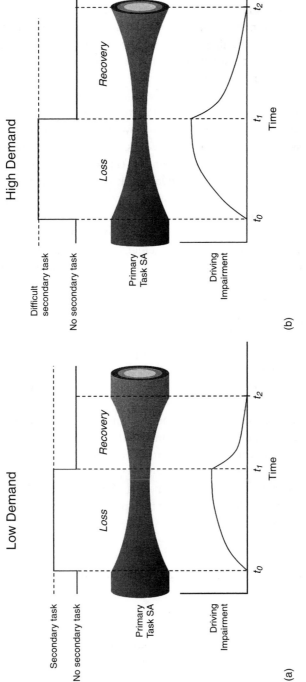

Note. Figure 6.3a illustrates a secondary task with low cognitive demand, an easier secondary task. Figure 6.3b illustrates a secondary task with high demand, a much harder secondary task. SA = situation awareness.

task. In Figures 6.3a and 6.3b, the secondary task was equivalent in task duration. The loss of situation awareness is greater, the recovery of situation awareness is slower, and the impairment to driving is larger when motorists perform a more demanding secondary task (see Figure 6.3b). The impact of a secondary task on situation awareness is represented by the decrease in the volume of the cylinder from t_0 to t_2. Generally speaking, secondary tasks that have a greater demand, have a longer duration, or both will have a more detrimental impact on a driver's situation awareness.

The pattern depicted in Figure 6.3 shows the interrelationship between primary-task performance, secondary-task demand, and situation awareness in the attention economy. Given the capacity limits of attention and driving, as the difficulty of a secondary task increases, situation awareness is lost, driver behavior becomes more reactive, and driving performance is degraded. Factors that modulate this relationship include the demands of the primary task of driving, nature and complexity of the secondary task, and expertise and arousal of the driver. Together, these factors impact the loss and recovery of situation awareness. The temporal dynamic suggests a hysteresis that persists after a discrete secondary-task event has been completed.

Returning to our example of the motorist using their voice to send a short text message, Figure 6.3a illustrates a well-designed system that places a relatively low demand on the driver when they use it. By contrast, Figure 6.3b illustrates a poorly designed system that places a relatively high demand on the driver. For illustrative purposes, the time to send the text message—the task duration—is held constant; however, with real-world systems, ease of use often covaries with task duration (i.e., systems that are more difficult to use often have longer task interaction times, e.g., Strayer et al., 2017). Figure 6.3 shows the relationship among the ease of use, driver's mental workload, and the resulting loss and recovery of situation awareness. Poorly designed systems result in higher levels of operator workload, a greater loss of situation awareness, and a longer period of recovery time. The difference between systems can be visualized as the volumetric difference between the left- and right-hand cylinders.

MEASUREMENT OF DYNAMIC FLUCTUATIONS IN THE ATTENTION ECONOMY

A promising new technique for assessing the dynamic fluctuations in workload such as those depicted in Figure 6.3 involves the use of the detection response task (DRT). The DRT is an International Standards Organization protocol that involves presenting a secondary stimulus (e.g., a changing light or vibrating buzzer) every 3 to 5 seconds and requiring the participant to respond to these events when they detect them by pressing a microswitch attached to their finger. As the workload increases, the reaction time to the DRT stimulus increases, and the likelihood of detection of the DRT stimulus (i.e., the hit rate) decreases (e.g., Strayer et al., 2013, 2015; Strayer, Cooper, et al., 2017).

The DRT protocol has proven to be sensitive to dynamic changes in the driver's workload and provides an objective assessment of the driver's workload associated with different interactions within the vehicle. To illustrate the potential of this technique, Strayer, Biondi, and Cooper (2017) used the DRT to simultaneously assess the workload of both a driver and a nondriver when the two were engaged in either a passenger conversation or a cell phone conversation. This method deployed two DRT devices that were yoked to measure the dynamic workload of each member of the conversational dyad when they were talking and when they were listening. With this yoked DRT procedure, an LED light mounted to a flexible arm connected to a headband was positioned in the periphery of left eye of each member of the dyad. The DRT devices (one for the driver and one for the nondriver) were synchronized so that the LED lights were presented simultaneously. Each of the participants responded separately to the light by pressing a microswitch attached to their finger. A microphone was also attached to each DRT device to identify if the driver or nondriver was talking at any given point in time.

The DRT data obtained from the driver and nondriver are presented in Figure 6.4. The "passenger" conditions reflect the conversation for the driver and nondriver when they are both in the vehicle. The "cell phone" conditions refer to the conversation for the driver and nondriver over a hands-free cell phone. The extension "DT" refers to situations in which the driver is talking, and the extension "DL" refers to situations in which the driver is listening (when the driver is talking, the passenger is listening, and vice versa). The single-task condition reflects performance when the driver is operating the vehicle and not talking to anyone.

The sawtooth pattern illustrated in Figure 6.4 shows the dynamic sensitivity of the DRT to workload as the driver and passenger conversed. The DRT data show that the workload for both the driver and passenger was greater when they were talking than when they were listening. Moreover, the pattern obtained with passenger conversations was the same as that obtained with cell phone conversations (for both the driver and the nondriver). In both cases, the reaction time to the DRT light was longer when the participant was talking than when they were listening. The driving task also added additional workload to the driver, as evidenced by the differences between the driver and the nondriving interlocutor.

FIGURE 6.4. Mean Reaction Time as a Function of Condition

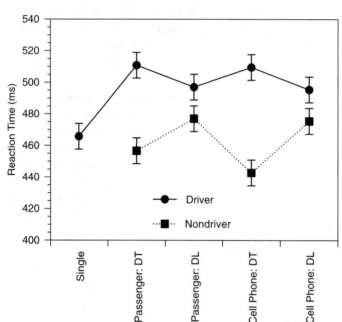

Note. Error bars reflect the standard error of the mean. The "passenger" conditions reflect the passenger conversation for the driver and nondriver. The "cell phone" conditions refer to the cell phone conversations for the driver and nondriver. DT refers to situations in which the driver is talking, and DL refers to situations in which the driver is listening. The single-task condition reflects performance when the driver is driving and not talking to anyone.

These data show that the DRT protocol is a useful metric for quantifying the dynamic ebb and flow of workload in the attention economy. As attention is diverted from driving by secondary-task engagement, the SPIDER-related processes are starved for resources, the situation awareness of the driver is diminished, and the impairments to driving increase. The DRT provides a method for uncovering both the loss and recovery of situation awareness as depicted in Figure 6.3 (e.g., Strayer et al., 2017).[3] Because of the temporal precision of the measure, the best estimates for the loss and recovery functions are provided by the DRT; however, other measures of driving performance have shown similar loss and recovery patterns (Jenness et al., 2015).

---

[3] For the DRT to be an effective metric for the dynamic allocation of attention, the onset of the DRT stimulus must be independent of the events in the driving environment. Consequently, the DRT will probe equally often throughout the t_0 to t_2 interval presented in Figure 6.3.

VEHICLE AUTOMATION

Vehicle automation promises to improve transportation efficiency and increased traffic safety. Recent estimates suggest that automated vehicle technology would reduce the number of road crashes and fatalities from 25% to 90% (European Commission, Directorate General for Transport, 2016; Litman, 2017). However, fully autonomous self-driving cars are in the distant future, despite cunning publicists who may tell you otherwise (e.g., Reuters, 2016). Today, semiautomated systems (i.e., level-2 and level-3 automation[4]) let drivers relinquish control of acceleration and steering operations to assistance systems (via the combined use of adaptive cruise control and lane keeping assist systems) with the assumption that the human driver will monitor the systems and be ready to resume manual control of driving if the autopilot fails (e.g., if the lane markings become worn). Nonetheless, such drastic shift in vehicle operation comes with challenges in the attention economy.

However, humans are notoriously poor at performing monitoring tasks. The literature on user interaction with automation in the maritime and aviation fields shows the ability to monitor the functioning of automated systems diminishes over time because of boredom, understimulation, and vigilance decrements (Cymek et al., 2016; Endsley & Kiris, 1995; Parasuraman & Manzey, 2010; Parasuraman et al., 1993; Singh et al., 1997; Wickens et al., 2015). The Yerkes-Dodson law (Yerkes & Dodson, 1908) suggests that there is an inverted U-shape relationship between arousal and performance such that extremely low or high levels of arousal will lead to a decline in performance. Applied to the context of driver interaction with automated systems, it is predicted that a reduction in overall levels of workload caused by the human-to-system transition of vehicle control will cause drivers to be understimulated and, as a result, they will gradually disengage from the driving and monitoring tasks.

Early studies on driver interaction with assistance systems and semi-automated vehicle technology show the potential for vehicle automation to push drivers out-of-the-loop and impoverish SPIDER-related processes. Vollrath et al. (2011), investigated effects produced by increasing levels

---

[4]The Society of Automotive Engineers (SAE On-Road Automated Driving Committee, 2014) defines level 2 and level 3 as semiautomated. In a level-2 semiautomated system, driver assistance systems are in charge of steering and acceleration/deceleration, and the human driver is expected to monitor the driving task and respond to a request by the system to intervene. In a level-3 semiautomated system, the automated driving system is responsible for driving and monitoring the driving environment. The human driver in expected to respond appropriately to a request to intervene.

of longitudinal control automation on driver workload and reactivity to changing road conditions. Participants drove a simulated vehicle in three conditions: manual driving, regular cruise control (the system controls the speed but not the distance to the forward vehicle), and adaptive cruise control (the system automatically controls the speed while maintaining a safe distance to the forward vehicle). Although having speed control systems engaged resulted in apparent safety benefits (e.g., average maintained speed was lower in both cruise control and adaptive cruise control conditions than during manual driving), the driver-to-system handover of vehicle longitudinal operations led to a decline in driver arousal and a decrement in their reactivity to changing road conditions. Stanton and Young (2005) found that driving with adaptive cruise control engaged caused drivers to lose awareness of the surrounding traffic. A similar phenomenon was also observed in the on-road study by Biondi et al. (2017) involving participants who drove a semiautomated vehicle on the road. Operating the lane keeping assist system (in which the system automatically maintains the vehicle within the lane without additional inputs from the driver) led motorists to gradually become disengaged from the driving and monitoring tasks after the 60-minute highway drive.

Another relevant issue within the context of user interaction with vehicle automation is trust—or absence thereof. Increased trust in or overreliance on automated systems has been found to increase the likelihood of users misusing or abusing automation. In the study by Reichenbach et al. (2010) on spacecraft control operators, an extended exposure to an apparently reliable system caused an increase in user trust in the system but, as a result, impoverished monitoring of its functioning and the detection of system failures (for similar results, see Bailey & Scerbo, 2007; Hoc, 2000). The decrement in vigilance and performance following the introduction of automated systems is commonly referred to as the *paradox of automation* (Hancock & Szalma, 2008). As drivers become more trusting and complacent with automated vehicles, their monitoring of the system declines, thus increasing the likelihood of promptly regaining control of the vehicle if a system failure occurs (e.g., Hergeth et al., 2016; Payre et al., 2015).

Using SPIDER, Figure 6.5 presents a scenario with a level-2 semiautonomous driving system in which a motorist activates the autonomous features in the vehicle at t_0. At t_1, the autonomous features of the vehicle disengage, and the driver is expected to take control of the vehicle. Between t_0 and t_1, there may be a loss of situation awareness (as depicted in Figure 6.5) associated with the driver's ceasing to monitor the driving environment. A critical challenge for semiautonomous vehicles is determining the level of driver engagement and their level of situation awareness.

FIGURE 6.5. The Hypothesized Effect of Level-2 Semiautonomous Systems on a Driver's Situation

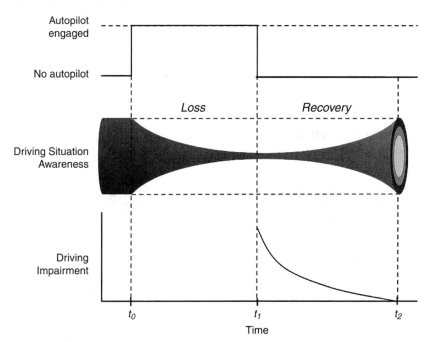

Note. Situation awareness is high at t_0, the moment that a motorist engages autopilot. The moment when autopilot becomes disengaged is reflected by t_1. Between t_0 and t_1, situation awareness is systematically lost. Between t_1 and t_2, situation awareness is slowly recovered. The timescale in Figures 6.2 and 6.5 are not the same. Notably, in Figure 6.5, the t_0-t_1 interval is likely to be much greater—on the order of minutes or even hours.

Vehicle manufactures have developed surrogate measures to encourage driver engagement, including the use of steering wheel torque. For example, lane keeping assistive systems often issue a visual hands-off warning, followed later by a deactivation of the lane keeping system, after prolonged hands-off driving. However, simply placing the hands on the steering wheel is often an inadequate surrogate for driver engagement and provides little information on where the driver is focusing their attention (National Transportation Safety Board [NTSB], 2016).[5] Because of the self-driving features of the vehicle, there is likely to be no impairment to driving when autopilot is

---

[5] A perusal of http://youtube.com reveals several unsafe methods that people have developed to circumvent the surrogate measures used to determine driver engagement.

engaged. However, at t_1, the driver is expected to take over control of the vehicle. The paradox of automation will lead drivers to become complacent in monitoring the driving environment, and, consequently, their situation awareness will be degraded. The interval between t_1 and t_2 represents the interval of time for the driver to recover situation awareness. Depending on the degree of situation awareness lost by t_1, the transition of control could be challenging for the driver.

A DEADLY WANDERING: CONSEQUENCES FOR SITUATION AWARENESS

The use of automation may have unintended consequences on drivers' situation awareness that can lead to tragic events. In May 2016, a Tesla Model S was traveling at 74 mph down a Florida highway under the control of the vehicle's adaptive cruise control and auto-steer functions when the vehicle collided with a semitruck broadside. The 40-year-old driver of the Tesla died in the crash, whereas the truck driver was unhurt. According to an investigation by the NTSB (2017), neither the Tesla's autonomous systems or the car's driver made an attempt to avoid the impact. The report indicated that both the Tesla design and the driver's behavior played a role in the degradation of situation awareness that led to the crash. The Tesla relies on steering wheel inputs to measure driver engagement, which the NTSB report deemed insufficient because it does not require visual engagement in the driving task to apply pressure to the steering wheel. On the other hand, the report also found that the driver likely did not understand the limitations of the vehicle's autopilot and, as a result, relied too heavily on automation in situations in which it is more likely to fail.

This tragic crash highlights the consequences of a loss of situation awareness and overconfidence in the attention economy. According to the National Highway Traffic Safety Administration (2016), the driver of the Tesla had 7 seconds during which he could have taken evasive actions (e.g., braking, steering) to avoid the crash. Seven seconds should be sufficient for an attentive driver to react to an imperative event; indeed, estimates assume that an attentive driver can initiate evasive action within 1.5 seconds of an imperative event (Olson & Sivak, 1986). However, an examination of the Tesla sensor data indicated that no evasive action was undertaken before the crash. The precise reason for why the driver failed to react is still unknown; however, driver inattention, overreliance on automation, and prolonged disengagement from the driving task are likely factors (NTSB, 2017). The postcrash investigation by the NTSB found that during the

Tesla driver's 41-minute trip, autopilot was active for 37 minutes, and the system detected driver-applied torque on the steering wheel for a total of only 25 seconds. This pattern of disengagement and inattention is depicted in Figure 6.5 as a loss of situation awareness with inadequate recovery resulting in an adverse event.

CONCLUSION

Identifying the factors that govern motorists' loss and recovery of situation awareness is critical to traffic safety. Ongoing research suggests that the complexity of the driving environment, complexity of the secondary-task interaction, and duration of the secondary task are critical components governing SPIDER and situation awareness in the attention economy. Innovative methods, such as the International Standards Organization standard DRT, provide a sensitive and dynamic metric for understanding the attention economy. However, much less is known about the attention economy and self-driving vehicles. Current versions of semiautomated vehicles change the nature of driving from a control task to a monitoring task. Humans are notoriously poor at performing monitoring tasks like those required with semiautonomous systems. The capacity limitations in the attention economy suggest that there will be unintended consequences of this emerging technology.

REFERENCES

Atchley, P., Tran, A. V., & Salehinejad, M. A. (2017). Constructing a publically available distracted driving database and research tool. *Accident Analysis & Prevention*, *99*(Part A), 306–311. https://doi.org/10.1016/j.aap.2016.12.005

Bailey, N. R., & Scerbo, M. W. (2007). Automation-induced complacency for monitoring highly reliable systems: The role of task complexity, system experience, and operator trust. *Theoretical Issues in Ergonomics Science*, *8*(4), 321–348. https://doi.org/10.1080/14639220500535301

Biondi, F., Goethe, R., Cooper, J., & Strayer, D. (2017). Partial-autonomous frenzy: Driving a level-2 vehicle on the open road. In D. Harris (Ed.), *Engineering psychology and cognitive ergonomics: Cognition and design* (EPCE 2017, Lecture Notes in Computer Science, Vol. 10276; pp. 329–338). Springer, Cham. https://doi.org/10.1007/978-3-319-58475-1_25

Biondi, F., Turrill, J. M., Coleman, J. R., Cooper, J. M., & Strayer, D. L. (2015, June 22–25). Cognitive distraction impairs driver's anticipatory glances: An on-road study. In D. V. McGehee, J. D. Lee, & M. Rizzo (Eds.), *Proceedings of the Eighth International Driving Symposium on Human Factors in Driver Assessment, Training, and Vehicle Design* [Symposium]. 2015 Driving Assessment Conference, Salt Lake City, UT, United States. https://doi.org/10.17077/drivingassessment.1546

Braver, T., Gray, J., & Burgess, G. (2007). Explaining the many varieties of working memory variation: Dual mechanisms of cognitive control. In A. R. A. Conway, C. Jarrold, M. J. Kane, A. Miyake, & J. N. Towse (Eds.), *Variation in working memory* (pp. 76–106). Oxford University Press.

Briggs, G. F., Hole, G. J., & Turner, J. A. (2017). The impact of attentional set and situation awareness on dual tasking driving performance. *Transportation Research Part F: Traffic Psychology and Behaviour, 57*, 36–47. https://doi.org/10.1016/j.trf.2017.08.007

Brown, T. L., Lee, J. D., & McGehee, D. V. (2001). Human performance models and rear-end collision avoidance algorithms. *Human Factors, 43*(3), 462–482. https://doi.org/10.1518/001872001775898250

Caird, J. K., Willness, C. R., Steel, P., & Scialfa, C. (2008). A meta-analysis of the effects of cell phones on driver performance. *Accident Analysis & Prevention, 40*(4), 1282–1293. https://doi.org/10.1016/j.aap.2008.01.009

Cooper, J. M., Medeiros-Ward, N., & Strayer, D. L. (2013). The impact of eye movements and cognitive workload on lateral position variability in driving. *Human Factors, 55*(5), 1001–1014. https://doi.org/10.1177/0018720813480177

Cooper, J. M., Vladisavljevic, I., Medeiros-Ward, N., Martin, P. T., & Strayer, D. L. (2009). An investigation of driver distraction near the tipping point of traffic flow stability. *Human Factors, 51*(2), 261–268. https://doi.org/10.1177/0018720809337503

Cooper, P. J., & Zheng, Y. (2002). Turning gap acceptance decision-making: The impact of driver distraction. *Journal of Safety Research, 33*(3), 321–335. https://doi.org/10.1016/S0022-4375(02)00029-4

Cymek, D. H., Jahn, S., & Manzey, D. H. (2016). Monitoring and cross-checking automation: Do four eyes see more than two? *Proceedings of the Human Factors and Ergonomics Society Annual Meeting, 60*(1), 143–147. https://doi.org/10.1177/1541931213601033

Durso, F., Rawson, K., & Girotto, S. (2007). Comprehension and situation awareness. In F. T. Durso, R. S. Nickerson, S. T. Dumais, S. Lewandowsky, & T. J. Perfect (Eds.), *Handbook of applied cognition* (2nd ed., pp. 163–193). John Wiley & Sons. https://doi.org/10.1002/9780470713181.ch7

Endsley, M. R. (1995). Towards a theory of situation awareness in dynamic systems. *Human Factors, 37*(1), 32–64. https://doi.org/10.1518/001872095779049543

Endsley, M. R. (2015). Situation awareness misconceptions and misunderstandings. *Journal of Cognitive Engineering and Decision Making, 9*(1), 4–32. https://doi.org/10.1177/1555343415572631

Endsley, M. R., & Kiris, E. O. (1995). The out-of-the loop performance problem and level of control in automation. *Human Factors, 37*(2), 381–394. https://doi.org/10.1518/001872095779064555

Engström, J., Johansson, E., & Östlund, J. (2005). Effects of visual and cognitive load in real and simulated motorway driving. *Transportation Research Part F: Traffic Psychology and Behaviour, 8*(2), 97–120. https://doi.org/10.1016/j.trf.2005.04.012

European Commission, Directorate General for Transport. (2016, November). *Advanced driver assistance systems.* https://ec.europa.eu/transport/road_safety/sites/roadsafety/files/ersosynthesis2016-adas15_en.pdf

Fisher, D. L., & Strayer, D. L. (2014). Modeling situation awareness and crash risk. *Annals of Advances in Automotive Medicine, 58*, 33–39. https://www.ncbi.nlm.nih.gov/pmc/articles/PMC4001668/

Hancock, P. A., & Szalma, J. L. (Eds.). (2008). *Performance under stress*. Ashgate.

Harbluk, J. L., Noy, Y. I., Trbovich, P. L., & Eizenman, M. (2007). An on-road assessment of cognitive distraction: Impacts on drivers' visual behavior and braking performance. *Accident Analysis & Prevention, 39*(2), 372–379. https://doi.org/10.1016/j.aap.2006.08.013

He, J., Becic, E., Lee, Y. C., & McCarley, J. S. (2011). Mind wandering behind the wheel: Performance and oculomotor correlates. *Human Factors, 53*(1), 13–21. https://doi.org/10.1177/0018720810391530

Hergeth, S., Lorenz, L., Vilimek, R., & Krems, J. F. (2016). Keep your scanners peeled: Gaze behavior as a measure of automation trust during highly automated driving. *Human Factors, 58*(3), 509–519. https://doi.org/10.1177/0018720815625744

Hoc, J.-M. (2000). From human–machine interaction to human–machine cooperation. *Ergonomics, 43*(7), 833–843. https://doi.org/10.1080/001401300409044

Horrey, W. J., & Wickens, C. D. (2006). Examining the impact of cell phone conversations on driving using meta-analytic techniques. *Human Factors, 48*(1), 196–205. https://doi.org/10.1518/001872006776412135

Horrey, W. J., Wickens, C. D., & Consalus, K. P. (2006). Modeling drivers' visual attention allocation while interacting with in-vehicle technologies. *Journal of Experimental Psychology: Applied, 12*(2), 67–78. https://doi.org/10.1037/1076-898X.12.2.67

Ishigami, Y., & Klein, R. M. (2009). Is a hands-free phone safer than a handheld phone? *Journal of Safety Research, 40*(2), 157–164. https://doi.org/10.1016/j.jsr.2009.02.006

Jenness, J. W., Baldwin, C., Chrysler, S., Lee, J. D., Manser, M., Cooper, J., Salvucci, D., Benedick, A., Huey, R., Robinson, E., Gonzales, C., Lewis, B., Eisert, J., Marshall, D., Lee, J., Morris, N., & Becic, E. (2015). *Connected vehicle DVI design research and distraction assessment* (HFCV Phase 2 NHTSA DTNH22-11-D-00237 Task Order No. 0001). National Highway Traffic Safety Administration.

Kahneman, D. (1973). *Attention and effort*. Prentice-Hall.

Kass, S. J., Cole, K. S., & Stanny, C. J. (2007). Effects of distraction and experience on situation awareness and simulated driving. *Transportation Research Part F: Traffic Psychology and Behaviour, 10*(4), 321–329. https://doi.org/10.1016/j.trf.2006.12.002

Litman, T. (2017). *Autonomous vehicle implementation predictions: Implications for transport planning*. Victoria Transport Policy Institute. https://orfe.princeton.edu/~alaink/SmartDrivingCars/PDFs/VIctoriaTransportAV_Predictionsavip.pdf

Liu, Y., Singh, S., & Subramanian, R. (2015, October). *Motor vehicle traffic crashes as a leading cause of death in the United States, 2010 and 2011* (Traffic Safety Facts Research Note; Report No. DOT HS 812 203). National Highway Traffic Safety Administration. https://crashstats.nhtsa.dot.gov/Api/Public/ViewPublication/812203

Lu, Z., Coster, X., & de Winter, J. (2017). How much time do drivers need to obtain situation awareness? A laboratory-based study of automated driving. *Applied Ergonomics, 60*, 293–304. https://doi.org/10.1016/j.apergo.2016.12.003

Mack, A., & Rock, I. (1998). *Inattentional blindness*. MIT Press. https://doi.org/10.7551/mitpress/3707.001.0001

Medeiros-Ward, N., Cooper, J. M., & Strayer, D. L. (2014). Hierarchical control and driving. *Journal of Experimental Psychology: General, 143*(3), 953–958. https://doi.org/10.1037/a0035097

National Highway Traffic Safety Administration. (2016). *ODI resume* (DOT Investigation No. PE 16-007). National Highway Traffic Safety Administration. https://static.nhtsa.gov/odi/inv/2016/INCLA-PE16007-7876.pdf

National Transportation Safety Board. (2016). *Preliminary report, highway HWY16FH018*. https://www.ntsb.gov/investigations/AccidentReports/Pages/HWY16FH018-preliminary.aspx

National Transportation Safety Board. (2017, September 12). *Collision between a car operating with automated vehicle control systems and a tractor-semitrailer truck, Williston, FL, May 7, 2016 (Report No. NTSB/HAR-17-XX).* https://www.ntsb.gov/news/events/Documents/2017-HWY16FH018-BMG-abstract.pdf

Navon, D., & Gopher, D. (1979). On the economy of the human-processing system. *Psychological Review, 86*(3), 214–255. https://doi.org/10.1037/0033-295X.86.3.214

Olson, P. L., & Sivak, M. (1986). Perception-response time to unexpected roadway hazards. *Human Factors, 28*(1), 91–96. https://doi.org/10.1177/001872088602800110

Parasuraman, R., & Manzey, D. H. (2010). Complacency and bias in human use of automation: An attentional integration. *Human Factors, 52*(3), 381–410. https://doi.org/10.1177/0018720810376055

Parasuraman, R., Molloy, R., & Singh, I. L. (1993). Performance consequences of automation-induced "complacency." *International Journal of Aviation Psychology, 3*(1), 1–23. https://doi.org/10.1207/s15327108ijap0301_1

Payre, W., Cestac, J., & Delhomme, P. (2015). Fully automated driving: Impact of trust and practice on manual control recovery. *Human Factors, 58*(2), 229–241. http://doi.org/10.1177/0018720815612319

Ratcliff, R., & Strayer, D. (2014). Modeling simple driving tasks with a one-boundary diffusion model. *Psychonomic Bulletin & Review, 21*, 577–589. https://doi.org/10.3758/s13423-013-0541-x

Readinger, W. O., Chatziastros, A., Cunningham, D. W., Bülthoff, H. H., & Cutting, J. E. (2002). Gaze-eccentricity effects on road position and steering. *Journal of Experimental Psychology: Applied, 8*(4), 247–258. https://doi.org/10.1037/1076-898X.8.4.247

Recarte, M. A., & Nunes, L. M. (2000). Effects of verbal and spatial-imagery tasks on eye fixations while driving. *Journal of Experimental Psychology: Applied, 6*(1), 31–43. https://doi.org/10.1037/1076-898X.6.1.31

Regan, M. A., Hallett, C., & Gordon, C. P. (2011). Driver distraction and driver inattention: Definition, relationship and taxonomy. *Accident Analysis & Prevention, 43*(5), 1771–1781. https://doi.org/10.1016/j.aap.2011.04.008

Regan, M. A., & Strayer, D. L. (2014). Towards an understanding of driver inattention: Taxonomy and theory. *Annals of Advances in Automotive Medicine, 58*, 5–14.

Reichenbach, J., Onnasch, L., & Manzey, D. (2010). Misuse of automation: The impact of system experience on complacency and automation bias in interaction with automated aids. *Proceedings of the Human Factors and Ergonomics Society Annual Meeting, 54*(4), 374–378. https://doi.org/10.1177/154193121005400422

Reimer, B. (2009). Impact of cognitive task complexity on drivers' visual tunneling. *Transportation Research Record, 2138*(1), 13–19. https://doi.org/10.3141/2138-03

Reimer, B., Mehler, B., Wang, Y., & Coughlin, J. F. (2012). A field study on the impact of variations in short-term memory demands on drivers' visual attention and driving performance across three age groups. *Human Factors, 54*(3), 454–468. https://doi.org/10.1177/0018720812437274

Reuters. (2016, July 29). Mercedes yanks ad touting self-driving car. *Fortune.* http://fortune.com/2016/07/29/mercedes-withdraws-us-ad-touting-self-driving-car/

Rogers, S. D., Kadar, E. E., & Costall, A. (2005). Gaze patterns in the visual control of straight-road driving and braking as a function of speed and expertise. *Ecological Psychology, 17*(1), 19–38. https://doi.org/10.1207/s15326969eco1701_2

SAE On-Road Automated Driving Committee. (2014, January 16). *Taxonomy and definitions for terms related to on-road motor vehicle automated driving systems* (SAE Standard No. J3016_201401). SAE International. https://doi.org/10.4271/J3016_201401

Simons, D. J., & Chabris, C. F. (1999). Gorillas in our midst: Sustained inattentional blindness for dynamic events. *Perception, 28*(9), 1059–1074. https://doi.org/10.1068/p281059

Singh, I. L., Molloy, R., & Parasuraman, R. (1997). Automation-induced monitoring inefficiency: Role of display location. *International Journal of Human-Computer Studies, 46*(1), 17–30. https://doi.org/10.1006/ijhc.1996.0081

Stanton, N. A., & Young, M. S. (2005). Driver behaviour with adaptive cruise control. *Ergonomics, 48*(10), 1294–1313. https://doi.org/10.1080/00140130500252990

Strayer, D. L. (2015). Is the technology in your car driving you to distraction? *Policy Insights from the Behavioral and Brain Sciences, 2*(1), 157–165. https://doi.org/10.1177/2372732215600885

Strayer, D. L., Biondi, F., & Cooper, J. M. (2017). Dynamic workload fluctuations in driver/non-driver conversational dyads. In D. V. McGehee, J. D. Lee, & M. Rizzo (Eds.), *Driving Assessment 2017: International Symposium on Human Factors in Driver Assessment, Training, and Vehicle Design* (pp. 362–367). Public Policy Center, University of Iowa.

Strayer, D. L., Cooper, J. M., & Drews, F. A. (2004). What do drivers fail to see when conversing on a cell phone. *Proceedings of the Human Factors and Ergonomics Society Annual Meeting, 48*(19), 2213–2217. https://doi.org/10.1177/154193120404801902

Strayer, D. L., Cooper, J. M., Goethe, R. M., McCarty, M. M., Getty, D., & Biondi, F. (2017). *Visual and cognitive demands of using in-vehicle infotainment systems.* AAA Foundation for Traffic Safety.

Strayer, D. L., Cooper, J. M., Turrill, J. M., Coleman, J. R., & Hopman, R. J. (2015). *The smartphone and the driver's cognitive workload: A comparison of Apple, Google, and Microsoft's intelligent personal assistants.* AAA Foundation for Traffic Safety.

Strayer, D. L., Cooper, J. M., Turrill, J., Coleman, J., Medeiros-Ward, N., & Biondi, F. (2013). *Measuring cognitive distraction in the automobile.* AAA Foundation for Traffic Safety.

Strayer, D. L., & Drews, F. A. (2007). Cell-phone–induced driver distraction. *Current Directions in Psychological Science, 16*(3), 128–131. https://doi.org/10.1111/j.1467-8721.2007.00489.x

Strayer, D. L., Drews, F. A., & Johnston, W. A. (2003). Cell phone-induced failures of visual attention during simulated driving. *Journal of Experimental Psychology: Applied, 9*(1), 23–32. https://doi.org/10.1037/1076-898X.9.1.23

Strayer, D. L., & Fisher, D. L. (2016). SPIDER: A framework for understanding driver distraction. *Human Factors, 58*(1), 5–12. https://doi.org/10.1177/0018720815619074

Strayer, D. L., & Johnston, W. A. (2001). Driven to distraction: Dual-task studies of simulated driving and conversing on a cellular telephone. *Psychological Science, 12*(6), 462–466. https://doi.org/10.1111/1467-9280.00386

Taylor, T., Roman, L., McFeaters, K., Romoser, M., Borowsky, A., Merritt, D. J., Pollatsek, A., Lee, J. D., & Fisher, D. L. (2015). *Cell phone conversations impede latent hazard anticipation while driving, with partial compensation by self-regulation in more complex driving scenarios.* Arbella Insurance Human Performance Lab, University of Massachusetts.

Tsai, Y. F., Viirre, E., Strychacz, C., Chase, B., & Jung, T. P. (2007). Task performance and eye activity: Predicting behavior relating to cognitive workload. *Aviation, Space, and Environmental Medicine, 78*(Suppl. 1), B176–B185.

Turrill, J., Coleman, J. R., Hopman, R. J., Cooper, J. M., & Strayer, D. L. (2016). The residual costs of multitasking: Causing trouble down the road. *Proceedings of the Human Factors and Ergonomics Society Annual Meeting, 60*(1), 1967–1970. https://doi.org/10.1177/1541931213601448

Victor, T. W., Harbluk, J. L., & Engström, J. A. (2005). Sensitivity of eye-movement measures to in-vehicle task difficulty. *Transportation Research Part F: Traffic Psychology and Behaviour, 8*(2), 167–190. https://doi.org/10.1016/j.trf.2005.04.014

Vollrath, M., Schleicher, S., & Gelau, C. (2011). The influence of cruise control and adaptive cruise control on driving behaviour—A driving simulator study. *Accident Analysis & Prevention, 43*(3), 1134–1139. https://doi.org/10.1016/j.aap.2010.12.023

Wickens, C. D., Clegg, B. A., Vieane, A. Z., & Sebok, A. L. (2015). Complacency and automation bias in the use of imperfect automation. *Human Factors, 57*(5), 728–739. https://doi.org/10.1177/0018720815581940

Wilson, M., Chattington, M., & Marple-Horvat, D. E. (2008). Eye movements drive steering: Reduced eye movement distribution impairs steering and driving performance. *Journal of Motor Behavior, 40*(3), 190–202. https://doi.org/10.3200/JMBR.40.3.190-202

Wright, T. J., Samuel, S., Borowsky, A., Zilberstein, S., & Fisher, D. L. (2016). Experienced drivers are quicker to achieve situation awareness than inexperienced drivers in situations of transfer of control within a Level 3 autonomous environment. *Proceedings of the Human Factors and Ergonomics Society Annual Meeting, 60*(1), 270–273. https://doi.org/10.1177/1541931213601062

Yerkes, R. M., & Dodson, J. D. (1908). The relation of strength of stimulus to rapidity of habit-formation. *Journal of Comparative Neurology, 18*(5), 459–482. https://doi.org/10.1002/cne.920180503

PART **III** GETTING AWAY
AND LOOKING
FORWARD

7

HOW NATURE HELPS REPLENISH OUR DEPLETED COGNITIVE RESERVES AND IMPROVES MOOD BY INCREASING ACTIVATION OF THE BRAIN'S DEFAULT MODE NETWORK

RACHEL J. HOPMAN, RUTH ANN ATCHLEY, PAUL ATCHLEY, AND DAVID L. STRAYER

We clearly have a love–hate relationship with technology, as is discussed throughout this volume. Researchers commonly frighten us with data that illustrate the remarkable number of hours that members of modern communities spend using media and technology. However, some readers are likely asking, "So what?" Technology and media are ubiquitous in our culture. A lack of access to technology is even seen as a critical issue of social inequity. In a recent Pew Research Center report titled "Digital Differences" (Zickuhr & Smith, 2012), the authors discussed how inadequate access to modern technological tools and high-speed internet are critical factors that are diminishing the educational success of lower income children. In light of this side of the discussion, one can ask, "What could possibly be disadvantageous about just simply engaging with the tools and pastimes of modern society?"

To explore this question more fully, the current chapter does not seek to reproduce the topics and data outlining the potential negative consequences of heavy technology usage as addressed elsewhere in this book. Instead, we explore research that provides evidence about the kinds of environments and experiences we are "giving up" when we spend almost half of our day engaged with technology. Specifically, we discuss the drastic cultural shift

https://doi.org/10.1037/0000208-008
Human Capacity in the Attention Economy, S. Lane and P. Atchley (Editors)

from spending some portion of our day in natural environments to spending almost all of our time in technology-rich, built environments. This conversation begins with a review of the seminal work of Rachel and Stephen Kaplan (2011; S. Kaplan, 1995), who have worked to develop the attention restoration theory (ART). The work of the Kaplans and others has been highly influential in promoting the discussion of how our environment can significantly impact our attentional resources and other cognitive processes.

This chapter begins by reviewing ART and discussing how this theory posits that certain qualities must be present in an environment for our environmental surrounding to act in a restorative manner and not continue to drain attentional resources. Next, we review a sample of experimental data supporting the idea that the environments we are giving up—such as the natural environment—have distinct benefits not just for our physical well-being but for our cognitive well-being. The goal is to provide insight into the possible underlying neurophysiological and neuroanatomical mechanisms that contribute to the findings being discussed. *Soft fascination*, defined as internal reflection or mind wandering, is a key principle of nature restoration theories. The neural mechanism associated with mind wandering is defined as the *default mode network* (DMN), a network of regions that increase activity while the brain is at rest. The DMN plays an important role in memory formation and future planning, and could be the neural mechanism underlying the process of restoration in nature. This chapter explores the history of the DMN as it relates to mind wandering and explains how the DMN likely is a key component for restoration in nature to occur.

WHAT IS ATTENTION RESTORATION THEORY?

According to S. Kaplan (1995), ART states that spending time in natural environments can restore depleted cognitive resources, such as attention. For restoration to occur, ART suggests voluntary attention is a finite resource depleted when completing cognitively demanding tasks (S. Kaplan & Berman, 2010). Voluntary attention requires top-down processing to concentrate on the stimuli while disengaging attention from other external or internal thoughts (Ghatan et al., 1998). Digital technology uses voluntary attention to suppress the surrounding environment and focus on the quickly changing incoming stimuli (Boksem et al., 2005). Depleting these attentional resources can lead to cognitive fatigue (James, 1892) and decreased cognitive performance (Corbetta & Shulman, 2002). Cognitive fatigue increases risk of human-related error, such as committing traffic accidents, forgetting to take medication, or sending an email to the wrong person. Cognitive resources that optimize attention need to refuel and be restored. Using involuntary attention can counteract the negative effects from cognitive depletion. Involuntary attention is exogenous in nature and captured by the environment through bottom-up processing (Escera et al., 1998). This effortless form of attention is not focused on a task but rather pulls attention toward unique and interesting stimuli in the environment. Involuntary attention is subcategorized into two groups: *hard fascination*, when the mind is externally focused; and soft fascination, when the mind is internally focused (Herzog et al., 2003). Involuntary attention, specifically soft fascination, is hypothesized to restore the depleted resources needed for voluntary attention by allowing voluntary attention to recover (James, 1892; S. Kaplan, 1995).

Decades of research has shown that nature provides a unique environment conducive for restoration of attentional processing and improved mood. *Nature environments*—those that lack manmade structures and instead contain features natural to the surrounding landscape—provide unique stimuli that influence attentional processing. ART states that restorative activities and environments have four particular qualities: being away, fascination, extent, and compatibility (S. Kaplan, 2001). The environment must have enough extent to engage the sensory system as well as be away from mental or physical distractions associated with a cognitively demanding environment. The environment must be compatible with an individual's preferences and allow for soft fascination to occur. The individual often experiences soft fascination as the ability to internally think without having to process attention-demanding information (S. Kaplan & Berman, 2010). Soft fascination (interestingly, much like mind wandering) engages an internal state of attention in part by removing awareness of external stimuli.

Although involuntary attention is necessary for restoration to occur (S. Kaplan & Berman, 2010), a countering hypothesis could argue that simply deactivating voluntary attention results in restorative benefits to cognition. Similar to a muscle, voluntary attention tires with persistent use (S. Kaplan & Berman, 2010). Resting that muscle could restore functionality and replenish the depleted resources. Therefore, downregulation of voluntary attention could engage the process of restoration rather than, or in addition to, upregulation of involuntary attention. Natural environments remove stimuli, such as a brightly lit screen, that induce voluntary attention but also create a space harmonious for engaging in soft fascination, thus allowing for involuntary attention (Hartig et al., 2001). More research is necessary to determine the neural process underlying restoration and to decipher the restorative differences between activation of involuntary attention and deactivation of voluntary attention.

EMPIRICAL EVIDENCE THAT NATURE EXPOSURE IMPACTS COGNITION

Wilson's (1984) *biophilia* suggests that all humans have an innate appreciation for nature related to the connection of nature with our well-being. However, more than 50% of the world lives in an urban environment (Ohly et al., 2016). Perceived restorative benefits of nature are shown to influence behavior and mood, and the perception of a restorative environment is generally described as a serene, natural, and quiet refuge. For example, Stigsdotter et al. (2017) found that natural forest areas with rich diversity of vegetation and open views are rated as optimal for restoration. Unpredictable and spacious environments correlated to higher self-reported creativity (van Rompay & Jol, 2016), and naturalist gardens without elements of structure were perceived more restorative compared with formal, structured gardens (Twedt et al., 2016). Likewise, perception of stress decreased when viewing pictures of natural greenspaces (Van den Berg et al., 2014), and self-reported measures of well-being and happiness increased after exposure to natural environments (Korpela et al., 2017). Even greenery in shopping centers improved the shopping experience (Rosenbaum et al., 2016), and sounds emulating natural environments in grocery stores increased the tendency to purchase healthier food (Spendrup et al., 2016). Relaxing spaces and activities can allow for some restoration to occur, but the aesthetics of nature increases perceived restoration of cognition and mood (S. Kaplan & Berman, 2010).

ART researchers have also explored more direct behavioral/cognitive effects of nature restoration. They found that with exposure to nature, performance on behavioral tasks measuring attention and cognition improved (Gidlow et al., 2016; Hartig et al., 2003; Li & Sullivan, 2016; Tennessen & Cimprich, 1995) but that walking through or viewing an urban or neutral setting did not show similar improvements. A variety of cognitive tests measuring attention, such as the Backward Digit Span Test (Wechsler, 1987) and Sustained Attention to Response Task (Robertson et al., 1997), have been used to determine restoration after exposure to nature. However, each of these tests measures different elements of cognitive performance, such as vigilance, working memory, and self-regulation. Therefore, average performance varies based on the task and exposure process (Ohly et al., 2016). Simple attentional processes, such as orienting and alerting, demand fewer cognitive resources, are less susceptible to fatigue, and therefore would not result in the same benefits from nature restoration as an attentionally demanding task. Likewise, attentionally demanding tasks, such as the

Operation Span Task (Turner & Engle, 1989), drain cognitive resources and therefore should show improvement from exposure to nature.

Similar to ART, the stress reduction theory suggests that exposure to natural environments reduces stress and negative thoughts (Ulrich, 1983; Ulrich et al., 1991). Although self-report supports decreased stress levels in a natural environment, heart rate variability (Annerstedt et al., 2013) and cortisol levels (Bowler et al., 2010) show inconsistent responses when exposed to nature. Although previous research found decreased heart rates (Laumann, Gärling, & Stormark, 2003) and decreased skin conductance levels (de Kort et al., 2006) when one was exposed to videos of nature, more recent research did not find significant changes in cortisol levels or heart rate variability when spending time in natural environments. Future research is needed to determine if exposure to natural environments reduces stress. Conversely, positive affect significantly increases when exposed to natural environments (Berto, 2005; Felsten, 2009; Herzog et al., 1997, 2003; Laumann et al., 2001). Exposure to nature may significantly decrease stress and improve affect, which subsequently improves attentional processing.

Previous research has also shown nature exposure can be therapeutic to improve behavior, stress, and mood. For example, children with attention-deficit/hyperactivity disorder (ADHD) concentrate better on a cognitive task after a 20-minute walk in nature compared with a walk in an urban or neutral setting (Faber Taylor & Kuo, 2011), and students viewing green landscapes while learning show lower stress responses as measured through skin conductance and heart rate variability (Li & Sullivan, 2016). Students with views of nature also show faster recovery of stress responses after taking an exam compared with students with a view of buildings or no windows (Li & Sullivan, 2016). However, teens (16–18 years) are less likely to appreciate nature because of the difference in psychological arousal and social needs; they have less of an appreciation for nature and are more likely to associate nature with "fear" and "discomfort" (Bixler & Floyd, 1997; 1999; Herzog et al., 2000). Teens prefer to spend recreational time using technology for social connections and entertainment because nature isolates teens from social interactions and discourages using technology while outside (Hills et al., 2007). Greenwood and Gatersleben (2016) found teenagers spending time outdoors with friends reported increased positive affect and greater heart rate variability—a signal of decreased stress—compared with spending time outdoors alone, on a cell phone, or inside with friends. Nature provides unique restorative benefits for populations that suffer from attention problems; therefore, more research is necessary to determine the generalizability and applicability of nature therapy.

Testing higher order cognitive skills in a natural environment is a challenge; however, our research team (Atchley et al., 2012, 2019) measured complex cognition in participants who were exposed to nature over sustained periods as well as for shorter exposure times. Consistent with ART, Atchley et al. (2012) found that 4 days of immersion in nature and the corresponding disconnection from multimedia and technology increase performance on the Remote Associates Test (Mednick, 1962), which is a creativity, problem-solving task. We found that performance on the test increased by a full 50% in a group of naive hikers. Likewise, in a second study we recently completed (Atchley et al., 2019), we found that for participants who spent 45 minutes to 1 hour in a natural setting, we also observed a reliable improvement on the Remote Associates Test. In contrast, the group of participants who had stayed within a built environment engaged with media showed a numerically small (statistically nonsignificant) decrease in their performance on the same creativity task. Interestingly, when we used sustained nature exposure, we saw about a 50% increase in verbal creativity, whereas in our second study, the improvement was closer to a 20% improvement. This was a statistically reliable positive change, and given that the nature "dosage" was so significantly reduced in the second study, this remains a striking finding. Additionally, this pair of findings might be taken to indicate a potential "dose–response" relationship between nature exposure and cognitive improvement that would be interesting to explore further. Another important point about this work on creativity is that, although we had in the first group an unavoidable confound of nature exposure and aerobic exercise (which research [e.g., Hillman et al., 2008] conclusively demonstrates also has significant cognitive benefits), we did not have this same confound in our second study. In the short-exposure study, the two groups (those spending time in nature—the outdoor group—and those spending time in a built environment—the inside group) were asked only to walk a short distance in the study protocol and did not engage in any kind of aerobic exercise.

Our team has begun to look at other kinds of higher order cognition, such as problem solving and interpersonal affiliation. We recently used a group problem solving task called the *change of work procedure* problem (first developed by Maier in 1958; Hoffman et al., 1962; Maier & Solem, 1962) in we which we judged if the outdoor group and inside group (a) selected one of the solutions that were provided to them in the experimental protocol or came up with a novel solution that was not provided to them, (b) rated the degree of consensus and cohesion achieved within the groups, (c) examined the interpersonal behaviors that occurred between participants as they engaged in solving the problem, and (d) examined the overall affective state of the participants (Atchley et al., 2019). Inconsistent with our experimental predictions, we did not find that the outdoor group came up with a more creative solution to the workplace problem. We observed that the two groups were statistically equal in both the complexity and creativity of the solutions offered. So, the prediction that the benefit from spending time in a natural environment would lead to more creative solutions in our workplace problem was not supported.

However, we (Atchley et al., 2019) did find that the outdoor group achieved reliably greater consensus in adopting the solution that they picked. Our results seem to provide evidence that spending time in nature does not lead the group to come to a more creative solution; however, group members do seem to work better as a team. Specifically, what we found is that the group members in the outdoor group reported having a different experience in the problem-solving task in three critical ways. First, the outdoor group members reported being reliably more satisfied overall with the decision that their group made and reported feeling they were having a significantly greater influence on the decision process. Second, the outdoor group members spent more time engaged in affiliative and cooperative behaviors than the indoor group. Results showed that, after their exposures, the outdoor group reported reduced negative affect.

Our assumptions underlying this experiment were that the free interplay of ideas and strong social cohesion present in the groups is fundamental to the development of high-quality solutions to problems. If group members feel they are effectively being heard and respected in the ideas that they contribute, this sense of efficacy will lead to a stronger outcome. To further test these ideas, we (Atchley et al., 2019) made video recordings of inside and outdoor group interactions with the hope that we would be able to observe and code behavioral differences between the groups that would explain observed differences in group problem solving and group cohesion. The behaviors that could be reliably observed and coded were helping behaviors (e.g., sharing writing instruments), verbal interruptions, laughter, looking around (and not at other team members), nodding in agreement, verbal disagreement, verbal agreement, and speaking. Some variables, such as helping, interrupting, and disagreement produced little data (fewer than eight observations per group on average). Other variables (looking around, nodding in agreement) failed to produce any reliable differences between the groups. However, the outdoor group was significantly higher on two variables in a way that was consistent with their indicators of the outdoor group's being more satisfied with their group problem solving task than members of the inside group. First, the outdoor group was reliably more interactive with more observed instances of speaking than the inside group, and the outdoor group laughed about twice as much as the inside group. We hypothesize that a critical component leading to this increase in effective group interactions is the increased emotional empathy and emotional sensitivity that becomes possible once we both disengage from media and technology and engage in spending time in natural settings. These conclusions are consistent with our finding that the outdoor group reported a reliable decrease in negative affect. These conclusions are also consistent with other recent work that has examined the impact of nature engagement (and technology disengagement) on interpersonal facility and prosociality (Uhls et al., 2014; Zhang et al., 2014).

Exposure to natural environments has also been found to significantly change neuropsychological responses that reflect cognitive processing. Previous research found an increase in neural signatures related to meditation and a decrease in neural signatures related to arousal when viewing pictures of natural settings (Roe et al., 2013). Likewise, Aspinall et al. (2015) found decreased levels of frustration and arousal when walking in nature compared with walking in an urban environment. In our previous research measuring the neurophysiological effects of cognitive restoration using electroencephalography, we found decreased posterior alpha power (8–12 Hz) after a prolonged period in a natural environment unplugged from technology. The alpha frequency in the posterior regions is commonly associated with attentional fluctuations (Bowman et al., 2017) and is negatively correlated with onset of attention (Laufs et al., 2003). Therefore, the decreased activation of the alpha frequency correlates with upregulation of involuntary attention. Likewise, using technology upregulates the demands of voluntary attention (Strayer et al., 2003). Participants who walked while engaging with technology showed an increase in midline frontal theta power—a neuroelectric signal of voluntary attention (Onton et al., 2005) and cognitive workload (Anguera et al., 2013; Wascher et al., 2014)— whereas participants without technology did not show significant changes in theta activity. Midline frontal theta and posterior alpha activation represents the change in the attention-related neural networks, thus reflecting the activation of involuntary attention when spending time in nature. This exploratory research is one of the first to determine neurophysiological correlates of restoration in nature.

Although most people cannot spend prolonged periods in nature, regular exposure to nature can restore cognitive functioning by offloading the demand of attention (Berman et al., 2008). ART research establishes the importance of maintaining access to natural environments, such as local parks and greenspaces, to benefit cognitive health (Felsten, 2009; Fuller et al., 2007). These studies also demonstrate the importance of spending time in nature without technology to fully acquire the benefits of attentional restoration (Kahn et al., 2008). Putting down the phone and stepping away from the computer are key elements to restore cognitive functioning. Likewise, ART research enhances scientific knowledge of the mechanisms underlying cognitive restoration by using multivariate methods to explore the benefits of nature (Berto, 2005; Felsten, 2009; Laumann et al., 2003). Future research can continue measuring neurophysiological activity in natural environments to further understand the process of cognitive depletion and restoration.

EXPLORATION OF THE DEFAULT MODE NETWORK

The brain at rest is not truly a quiet brain. When individuals are not engaged in a task, a unique network of brain regions activates. Buckner and Vincent (2007) were the first to coin the name *default mode network*. The DMN is the most consistent pattern of activation found in functional magnetic resonance imaging (fMRI) research but has only been studied for the past 2 decades. Originally used as baseline brain activity, the DMN is defined by a network of neural regions that increases in activation when the brain is at rest. Once this unique pattern of activation was discovered, researchers speculated at the underlying processes at work. Many agree that the DMN is a preexisting continuous disposition of the brain (Raichle & Snyder, 2007), and the onset of activation in the DMN correlates with a decrease of focused attention (Buckner et al., 2008). However, the purpose of the DMN activation in relation to attention and memory remains debated. The DMN is activated when we are in a state of internal focused thought, specifically about self-referential future-oriented plans (Smallwood et al., 2012). The DMN is most commonly associated with *undirected thoughts*, defined as mind wandering (Andrews-Hanna, Reidler, Huang, & Buckner, 2010). Differences in internalized thoughts at rest, such as mind wandering, mindfulness, and soft fascinations, provide insight to determine the cognitive processes underlying the activation of the DMN. The DMN is proposed to be an underlying neural network related to attention restoration from nature (S. Kaplan & Berman, 2010).

Anatomically, the DMN is connected through an organized network of regions that can be detected in fMRI data at low frequencies (0.01–0.05 Hz) during rest (Damoiseaux et al., 2006; De Luca et al., 2006). The activation of the DMN during rest is one of the most consistent networks found in fMRI literature (Andrews-Hanna, Reidler, Sepulcre, et al., 2010). The DMN consists of gray matter regions connected by white matter tracts that control the onset and offset of low-frequency activations. Although previous research suggests that white matter tracts are not active during DMN activation (Fransson, 2005), more recent research has found evidence of activation using diffusion tensor imaging to measure white matter tracts, such as the cingulate bundle, that connect the DMN regions (Teipel et al., 2010; van den Heuvel & Hulshoff Pol, 2010). These tracts link key regions of the DMN, such as the medial prefrontal cortex (PFC), posterior cingulate cortex (PCC), left and right medial temporal lobes (MTL), lateral temporal cortex, retrosplenial cortex, inferior parietal lobule, and precuneus and hippocampal regions (see Figure 7.1; Buckner et al., 2008; Fransson & Marrelec, 2008). The DMN comprises two subsystems connected by the PCC and the anterior medial PFC (Andrews-Hanna, Reidler, Huang, & Buckner, 2010). The dorsal medial PFC system (comprising the dorsal medial PFC, lateral temporal cortex, temporal pole, and temporoparietal junction) and the MTL system (comprising the ventral medial PFC, posterior inferior parietal lobule, retrosplenial cortex, and hippocampal formation) have different functionality. According to Andrews-Hanna, Reidler, Huang, and Buckner (2010), the *dorsal medial PFC system* is engaged when people are making self-relevant decisions, whereas the *MTL system* is activated when these decisions require a mental model based on episodic memory (see Figure 7.2). Researchers have also found activation in the superior parietal and superior temporal cortex, anterior caudate precuneus, and left para-central lobule as well as the sensory cortex, which is involved with higher order processing (Damoiseaux et al., 2006; Mazoyer et al., 2001). In general, the key neural regions of the DMN are associated with self-reflection, mind wandering, and emotional processing (Fransson, 2006), which, one could argue, indicate the underlying purpose of the DMN.

WHAT IS THE PURPOSE OF THE DEFAULT MODE NETWORK?

Researchers have theorized that the brain has two proposed modes of internal processing: mental explorations and focused information extraction. These processes describe activation of the DMN—a neural network associated with internal processing and resting states of attention—and the "task positive" *attention network* (AN; Buckner et al., 2008; Fox et al., 2005), a neural network associated with engagement of voluntary attention. Although the AN and DMN can be activated simultaneously, these two networks are mostly anticorrelated (Christoff et al., 2009; Vincent et al., 2007). The ventral

FIGURE 7.1. The Brain's Default Mode Network (DMN) Comprises Several Subsystems Across the Cortex as Well as Subcortical Regions, Including the Hippocampal Formation

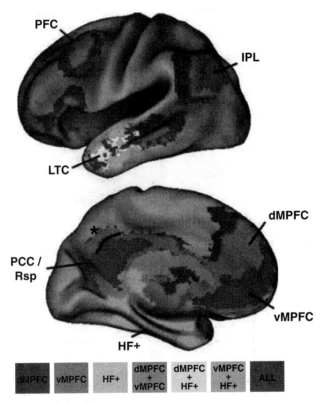

Note. Top panel: DMN substructures mapped onto the left cortical hemisphere. Bottom panel: DMN substructures mapped onto the right cortical hemisphere. PFC = prefrontal cortex; IPL = inferior parietal lobule; LTC = lateral temporal cortex; PCC/Rsp = posterior cingulate cortex/retrosplenial cortex; dMPFC = dorsal medial prefrontal cortex; vMPFC = ventromedial prefrontal cortex; HF+ = hippocampal formation, including a portion of the parahippocampal cortex. From "The Brain's Default Network: Anatomy, Function, and Relevance to Disease," by R. L. Buckner, J. R. Andrews-Hanna, and D. L. Schacter, 2008, *Annals of the New York Academy of Sciences, 1124*, p. 12 (https://doi.org/10.1196/annals.1440.011). Copyright 2008 by John Wiley & Sons. Reprinted with permission.

medial PFC, PCC, and other regions of the DMN negatively correlate to the AN, which is defined by a network of regions, including the posterior parietal cortex, inferior temporal gyrus, and anterior cingulate cortex (Smallwood et al., 2012). These opposing systems suggest a dichotomy between introspective and extrospective tasks and thoughts (Uddin et al., 2009). Better coordination between the DMN and AN leads to enhanced overall cognitive

FIGURE 7.2. The Posterior Cingulate Cortex and Anterior Medial Prefrontal Cortex Compose the Core Regions of the Default Mode Network and Connect the Two Underlying Subsystems

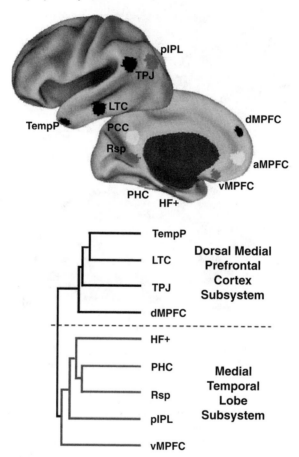

Note. The dorsal medial prefrontal cortex subsystem (shown in black) and the medial temporal lobe subsystem (shown in gray) have unique roles in relation to default mode network activity. pIPL = posterior inferior parietal lobe; TPJ = temporoparietal junction; LTC = lateral temporal cortex; TempP = temporal pole; PCC = posterior cingulate cortex; Rsp = retrosplenial cortex; dMPFC = dorsal medial prefrontal cortex; aMPFC = anterior medial prefrontal cortex; vMPFC = ventromedial prefrontal cortex; HF+ = hippocampal formation, including a portion of the parahippocampal cortex; PHC = parahippocampal cortex. From "Functional-Anatomic Fractionation of the Brain's Default Network," by J. R. Andrews-Hanna, J. S. Reidler, J. Sepulcre, R. Poulin, and R. L. Buckner, 2010, *Neuron, 65*(4), p. 552 (https://doi.org/10.1152/jn.00830.2009). Copyright 2010 by Elsevier Inc. Reprinted with permission.

performance. For example, participants with high working memory performance show a greater negative correlation between the DMN and the AN compared with low working memory performance (Hampson et al., 2006). Hampson and colleagues (2006) attributed this negative correlation to better coordinated activity between the PCC and medial PFC, thus suggesting improved synchronization of the onset of the DMN and offset of the AN. Thus, this idea emerges: Better coordination between these networks leads to enhanced overall cognitive performance.

Good coordination does not always mean that the systems will be "turned on" at different times in a reciprocal fashion. The brain is reflexive and driven by the demands of the environment (Sherrington, 1906); therefore, the DMN and AN adjust to accommodate the situation. For example, during novel attention-demanding tasks, both the DMN and AN are engaged. However, the DMN and AN do show fluctuating activation when correctly responding to an easy task requiring sustained attention (Christoff et al., 2009). Likewise, failure to suppress the posterior node of the DMN during a task results in a lapse of attention. Errors on a go–nogo task have been associated with an increase in DMN activity before the error, suggesting participants were inattentive to the task (Liang, Zou, He, & Yang, 2016). Using a 7-point Likert scale ranging from 1 (*completely on task*) to 7 (*completely off task*) for one question and ranging from 1 (*completely aware*) to 7 (*completely unaware*) for another, participants were randomly probed to report awareness to the sustained attention to response task (Christoff et al., 2009). The fMRI data analyzed just before the probe showed an increase in DMN activity for those who reported being unaware of the task and those who committed an error. Improper coordination between the DMN and AN resulted in performance failures on tasks measuring attention. However, cognitive restoration allowed both networks to optimize performance and flexibly switch, when necessary. Thus, restoration in nature could enhance neural flexibility to better perform tasks and better regulate the push–pull relationship of the DMN and AN.

ATTENTION RESTORATION THEORY AND DEFAULT MODE NETWORK: CONNECTING THE TWO

As the reader has probably already anticipated, the distinction of voluntary and involuntary attention in ART literature can be seen to parallel the opposing activation of the AN and the DMN. Voluntary attention activates regions associated with the AN, such as the anterior cingulate cortex and posterior parietal cortex (Greicius et al., 2003), whereas involuntary attention likely activates regions of the DMN. These opposing definitions of attention define two neural structures that represent the onset and offset of focused attention. Therefore, the functionality associated with each attention type reflects the functionality of the associated network (Corbetta & Shulman, 2002). According to S. Kaplan (1995), involuntary attention is required for restoration to occur. Consequently, it seems reasonable to predict that the process of attention restoration discussed in ART relates to increased activation of the DMN. However, more research is necessary to confirm that the activation of the DMN underlies the process of restoration in nature.

In 1931, Hans Berger noticed the brain is not idle when undirected, and the behavioral phenomena he observed has come to be called *mind wandering*, when one drifts into an internal state of attention and removes awareness of the external environment (Berger, 1931). Researchers have associated mind wandering with the DMN because disengagement from an attention-related task is necessary for the DMN to activate. However, other networks, such as the temporal memory network and the executive system, are also active when one is mind wandering based on relevance to the train of thought (Christoff, 2012). Thoughts independent of a particular task give a break to focused attention and are helpful for insight into task-related problem solving.

The activation of the DMN via positive, reflexive mind wandering is thought to have a restorative effect on attention, creativity, and overall mood (Baird et al., 2012; Smallwood et al., 2008). These specific areas of improvement are strikingly comparable with specific aspects of behavior and cognition that show enhancement following exposure to nature. Research has demonstrated that engaging in mind wandering facilitates creative problem solving through the process of incubation (Zedelius & Schooler, 2015). For example, after mind wandering, participants scored higher at the Unusual Uses Task (Baird et al., 2012) in which they produce as many unusual uses for an object as possible and also showed improved performance on a goal-directed task (Sio & Ormerod, 2009). Similarly, mind wandering improves creative and unique solutions on a task that requires problem solving (Förster et al., 2004) and increases positive affect (Hasenkamp et al., 2012). Mind wandering can engage creative thinking by disengaging focus from a task that is analogous to how activation of the DMN correlates to the disengagement of the AN. More specific to the focus of this chapter, mind wandering in nature shows improvements in cognitive functioning and well-being (Felsten, 2009; S. Kaplan & Berman, 2010).

Mind wandering during tasks can also be costly in that it results in deficits to performance on cognitive tasks that require external attention, such as reading comprehension (Schooler et al., 2011). Electroencephalography time-related waveforms associated with attention, motor cortices, and sensory cortices decreased when engaged in mind wandering as compared with on-task thoughts (Smallwood et al., 2008). Therefore, engaging in mind wandering lessens the ability to process information regarding the external environment. Decreased awareness of the external environment could result in dangerous consequences, such as inattentional blindness while driving (Strayer & Drews, 2007). Mason and colleagues (2007) found that mind wandering continues during easy and familiar tasks, and it decreases as tasks become less familiar or more difficult. Although in most cases, the AN downregulates as the DMN activates, Christoff and colleagues (2009) found that sudden, random mind wandering during a demanding task can activate both the AN and DMN. Christoff et al. (2009) proposed that these systems are mutually exclusive but potentially work in unison as infrequent stimulus-independent thoughts emerge during engagement with a task. Losing awareness of the external environment might be necessary for cognitive resources to replenish, but loss of awareness during an important attention-demanding task could have consequential results. Strategies such as mindfulness training or biofeedback could maximize benefits and reduce chances of improper DMN–AN regulation.

Mind wandering could have many potential functions, such as engaging in future planning and navigating social interactions as well as replenishing cognitive resources by providing a break from focused attention. However, increased daily use of technology prevents the mind from shutting off. Brightly lit screens and flashing signals keep attention alerted and deplete the AN, causing the DMN to activate during inappropriate times. Zoning out when trying to take a test or make a quick decision is the brain's cry for help. Allowing the AN to rest and the DMN to upregulate in natural settings that do not overload attention potentially drives the process of restoration.

SOFT FASCINATION: THE MIND'S INATTENTION TOWARD THOUGHTS

As mentioned earlier, soft fascination, a key component for restoration, is the ability to attend without exerting cognitive control or using attentional resources (S. Kaplan & Berman, 2010). Like mind wandering, soft fascination can be characterized by the tendency to drift into a state of thinking and reduce focused attention toward the external environment (Baird et al., 2012). Both mind wandering and soft fascination are defined by spontaneous, task-unrelated thoughts, or daydreaming (Christoff, 2012; Tennessen & Cimprich, 1995). However, mind wandering encompasses multiple types of internalized processing, including future planning; rumination on previous events; focused mental extractions; and nonplanned, sporadic thoughts. Thinking of a specific song lyric, making a mental checklist, and randomly remembering a specific event are all types of mental mind wandering. However, being absorbed by pleasant scenery and having no specific train of thought can also be defined as mind wandering. Ergo, soft fascination in natural environments should likely be defined as a subset of mind wandering. Soft fascination in nature allows for restoration to occur, just as activation of the MTL, a primary region of the DMN, can be triggered by mind wandering (Christoff et al., 2016; Mittner et al., 2016) and facilitates the recovery from mental fatigue (R. Kaplan & Kaplan, 2011; Zedelius & Schooler, 2015).

Natural environments present a quiet setting compatible for soft fascination to occur, and soft fascination allows for the resting brain to engage in a process of neural restoration. The purpose of soft fascination is likely to restore neural resources that fuel attentional control. Although the resources restored by nature exposure have not yet been defined, previous research has hypothesized that the attentional resources are sustained by the transmittance of dopamine (Nieoullon, 2002; Roiser et al., 2007). Engaging the AN during a cognitively demanding task depletes dopamine levels, and the amount of dopamine can modify attentional performance (Mattay et al., 2000). Research also has suggested that glucose stores can modulate attention (Benton et al., 1994; Wesnes et al., 2003). A depletion of glucose decreases cognitive performance (Kennedy & Scholey, 2000) and alters perception of the external environment (Schnall et al., 2010). Glucose is refreshed in other restorative activities, such as sleep (Van Cauter et al., 1991, 1997) and eating (R. J. Kaplan et al., 2000); therefore, cognitive restoration likely results in part through the restoration of glucose and dopamine. Future research is necessary to determine how exposure to nature modulates these neurophysiological resources, which are resources attributed to voluntary attention. Regardless of the neurophysiological mechanisms that allow for cognitive restoration, ART hypothesizes that resources depleted through voluntary attention are regained when spending time in nature (S. Kaplan, 1995; S. Kaplan & Berman, 2010). Nature creates an environment compatible for restoration to occur by removing cognitive distractors and providing an immersive setting that allows for soft fascination. During the process of restoration, the resting brain engages substratal mechanisms that initiate restoration of attention in the conscious mind, and these mechanisms are hypothetically explained by the onset of DMN and the offset of the AN.

Although it is easy to see how one can theorize a strong connection between the construct of soft fascination and the behavioral phenomena of mind wandering, one should also be aware of a theoretically and empirically supported relationship between soft fascination and mindfulness. *Mindfulness* is a mental state attained by intentionally focusing on the present environment (Cullen, 2011). Rather than internally reflecting on thoughts relevant to oneself, mindfulness meditation training encourages attention to focus on elements of the external environment (Eberth & Sedlmeier, 2012). Research shows mindfulness meditation also improves well-being (Brown & Ryan, 2003; Lutz et al., 2014; Meiklejohn et al., 2012) and decreases symptoms of depression (Ramel et al., 2004). Much like soft fascination, this type of training is thought to enhance attentional control (Cullen, 2011), and individuals show greater improvements when practicing mindfulness methods in natural environments (Lymeus et al., 2017). In contrast to mind wandering, mindfulness meditation requires effort to disengage with the stream of consciousness and focus on aspects in the surrounding environment. Both the medial PFC—associated with the DMN—and dorsolateral PFC— associated with the AN—activate during mindfulness meditation (Creswell et al., 2007). Before the DMN activates during mindful meditation, the AN is required to engage in the practice of mindfulness. Researchers have hypothesized that mindfulness meditation is similar to the experience of soft fascination and also might be mechanistically related to the process of restoration in nature (see, e.g., Lymeus et al., 2018). It is necessarily difficult— without the acquisition of more neuropsychological data—to determine if the primary mental state that results in attention restoration during the experience of soft fascination is engagement in mind wandering or engagement in mindfulness. It is also possible that both are beneficial, that the ideal mental state that can occur in nature varies between individuals, or both. Future research that explores these interesting and important issues is needed.

CONCLUSION AND FUTURE DIRECTIONS

As with many recently formed fields of study, many questions surrounding ART and the DMN literature are left to be answered by further research. More research is necessary to verify the association of the DMN with the process of nature restoration. The DMN is most commonly studied in fMRI literature; however, nature restoration cannot be best measured in a controlled laboratory setting. Although viewing pictures of nature is perceived more restorative than viewing pictures of other environments, perceived restoration likely does not parallel the restorative processes available in nature. We wish to argue that the neural mechanisms proposed through ART should be studied in situ to determine how the process of restoration occurs. When we measure neural mechanisms in the field, future research can determine physiological and neurophysiological correlates that equate to the onset of the DMN. For example, neuroelectric biomarkers of attention such as changes in resting alpha and theta frequencies might be one fruitful approach, given that changes in neural frequencies reflect activations and deactivations in neural networks. Likewise, further research might determine if other oscillatory frequencies correlate with the onset of the DMN.

Future studies are also necessary to determine if we can discern differences in creative performance along with other kinds of cognitive improvement that can be measured after mind wandering, engaging in mindfulness meditation, and sitting in nature. Previous studies have shown that each of these activities can enhance cognition and mood (Jang et al., 2011; R. Kaplan & Kaplan, 2011); however, more research is necessary to compare the restorative benefits among these three interrelated experiences. Likewise, research has yet to discover how cognitive performance changes over the course of a prolonged period in nature as well as to discern how long these restorative benefits last. Our own research shows a kind of dose–response effect, such that more time in nature leads to greater benefit. But this is just for one kind of cognitive performance. Also, we have used durations of nature exposure that are quite different (i.e., 4 days vs. 1 hour); clearly, much more could be done to map out the relationship between nature exposure duration and the benefits that can be gained.

Clinical populations, such as those with depression or attention-related issues, show tonic decreased activation of the DMN. However, research has yet to determine how these populations respond to nature. Future studies could measure the effects of nature restoration in specific high-stress clinical populations, such as those with posttraumatic stress disorder or ADHD. Nature-based therapy is currently used to treat symptoms of posttraumatic stress disorder (Rizzo et al., 2014; Sahlin et al., 2012; Wagenfeld et al., 2013) and ADHD (Safren et al., 2005); however, ART research has yet to determine if the process of restoration is different in these clinical groups. Nature therapy has a promising future in the field of psychology, but more research is necessary to determine the validity and generalizability of the cognitive effects of restoration. In addition, research needs to effectively explore the issue of the ideal "dose" of nature for a clinical population. All people, regardless of the degree of well-being, tend to have a limited amount of time to be outside. It is critical to have reliable data about how much nature exposure is needed to lead to an effective elevation of mood or cognitive resources.

Future studies could also compare the neurophysiological effects of virtual natural environments to restoration in "true" nature. Virtual reality (VR) has created the ability to apply a real-world scenario in a controlled laboratory setting. As technology becomes more advanced, VR can simulate environments that induce credible physiological and behavioral reactions (Sanchez-Vives & Slater, 2005). VR has been used as a therapy technique to expose patients to various real-life situations (Rothbaum et al., 1995) as well as used to measure perception (Interrante et al., 2006; Willemsen & Gooch, 2002), attention (Drews et al., 2008), and cognition (Kandalaft et al., 2013). Previous research has measured the effects of restoration in a low-grade nature VR setting compared with an immersive-nature VR setting (de Kort et al., 2006). The more immersive environment induced a stronger physiological reaction compared with the less immersive environment, suggesting greater restorative effects for the more immersive environment. Similarly, exposure to interactive VR nature decreased stress more than when one viewed neutral VR environments (Valtchanov et al., 2010). Current research efforts are using VR to quantify cognitive outcomes after nature exposure with and without physical activity. Empirically, ART is commonly measured using nonimmersive computer screens with pictures of different nature scenes (Berto et al., 2010; Felsten, 2009; Laumann et al., 2001). Although previous research has found significant outcomes measuring levels of restoration with these methods, research has yet to document how the type and level of immersion influences the restorative value. If a VR nature environment has restorative qualities, this form of exposure has many applications. Similar to nature therapy, nature-based VR exposure could also be used as therapy for specific clinical populations, such as those with limited access to nature. VR has the potential for scientists to quantify the benefits of nature by recreating the natural world in a sterilized environment and bringing the natural world into therapeutic settings.

In summary, exposure to natural environments positively affects health and cognitive functioning. However, the neural mechanisms underlying the process of restoration are yet to be determined in the ART literature. For restoration in nature to occur, S. Kaplan (1995) proposed that the natural environment must allow for soft fascination. Similar to soft fascination, mind wandering is the measurable behavioral component associated with the onset of the DMN. Therefore, the DMN likely activates when the mind is in a state of nonfocused, internal reflection, such as when spending time in nature. This chapter explored the history of the DMN and ART literature, outlined some of the research that establishes a connection between nature exposure and cognitive benefit, and investigated a potential connection between the mechanisms of the "resting" brain and the characteristics of nature that are the focus of ART. Although further research is necessary to determine the true functionality of the DMN, we would argue that the process of cognitive restoration likely is related to activation of the DMN. Natural environments improve well-being, but more research is necessary to determine how the process of restoration occurs.

REFERENCES

Andrews-Hanna, J. R., Reidler, J. S., Huang, C., & Buckner, R. L. (2010). Evidence for the default network's role in spontaneous cognition. *Journal of Neurophysiology, 104*(1), 322–335. https://doi.org/10.1152/jn.00830.2009

Andrews-Hanna, J. R., Reidler, J. S., Sepulcre, J., Poulin, R., & Buckner, R. L. (2010). Functional-anatomic fractionation of the brain's default network. *Neuron, 65*(4), 550–562. https://doi.org/10.1016/j.neuron.2010.02.005

Anguera, J. A., Boccanfuso, J., Rintoul, J. L., Al-Hashimi, O., Faraji, F., Janowich, J., Kong, E., Larraburo, Y., Rolle, C., Johnston, E., & Gazzaley, A. (2013). Video game training enhances cognitive control in older adults. *Nature, 501*, 97–101. https://doi.org/10.1038/nature12486

Annerstedt, M., Jönsson, P., Wallergård, M., Johansson, G., Karlson, B., Grahn, P., Hansen, Å. M., & Währborg, P. (2013). Inducing physiological stress recovery with sounds of nature in a virtual reality forest—Results from a pilot study. *Physiology & Behavior, 118*, 240–250. https://doi.org/10.1016/j.physbeh.2013.05.023

Aspinall, P., Mavros, P., Coyne, R., & Roe, J. (2015). The urban brain: Analysing outdoor physical activity with mobile EEG. *British Journal of Sports Medicine, 49*(4), 272–276. https://doi.org/10.1136/bjsports-2012-091877

Atchley, P., Atchley, R. A., Benau, E., & Strayer, D. (2019, October 1). *Building healthy teams: Using nature-based interventions to improve workplace performance and promote healthy work environments*. Shift. https://shiftjh.org/building-healthy-teams-using-nature-based-interventions-to-improve-workplace-performance-and-promote-healthy-work-environments/

Atchley, R. A., Strayer, D. L., & Atchley, P. (2012). Creativity in the wild: Improving creative reasoning through immersion in natural settings. *PLOS ONE, 7*(12), e51474. https://doi.org/10.1371/journal.pone.0051474

Baird, B., Smallwood, J., Mrazek, M. D., Kam, J. W., Franklin, M. S., & Schooler, J. W. (2012). Inspired by distraction: Mind wandering facilitates creative incubation. *Psychological Science, 23*(10), 1117–1122. https://doi.org/10.1177/0956797612446024

Benton, D., Owens, D. S., & Parker, P. Y. (1994). Blood glucose influences memory and attention in young adults. *Neuropsychologia, 32*(5), 595–607. https://doi.org/10.1016/0028-3932(94)90147-3

Berger, H. (1931). On the electroencephalogram of man. *European Archives of Psychiatry and Clinical Neuroscience, 94*(1), 16–60.

Berman, M. G., Jonides, J., & Kaplan, S. (2008). The cognitive benefits of interacting with nature. *Psychological Science, 19*(12), 1207–1212. https://doi.org/10.1111/j.1467-9280.2008.02225.x

Berto, R. (2005). Exposure to restorative environments helps restore attentional capacity. *Journal of Environmental Psychology, 25*(3), 249–259. https://doi.org/10.1016/j.jenvp.2005.07.001

Berto, R., Baroni, M. R., Zainaghi, A., & Bettella, S. (2010). An exploratory study of the effect of high and low fascination environments on attentional fatigue. *Journal of Environmental Psychology, 30*(4), 494–500. https://doi.org/10.1016/j.jenvp.2009.12.002

Bixler, R. D., & Floyd, M. F. (1997). Nature is scary, disgusting, and uncomfortable. *Environment and Behavior, 29*(4), 443–467. https://doi.org/10.1177/001391659702900401

Bixler, R. D., & Floyd, M. F. (1999). Hands on or hands off? Disgust sensitivity and preference for environmental education activities. *Journal of Environmental Education, 30*(3), 4–11. https://doi.org/10.1080/00958969909601871

Boksem, M. A., Meijman, T. F., & Lorist, M. M. (2005). Effects of mental fatigue on attention: An ERP study. *Cognitive Brain Research, 25*(1), 107–116. https://doi.org/10.1016/j.cogbrainres.2005.04.011

Bowler, D. E., Buyung-Ali, L. M., Knight, T. M., & Pullin, A. S. (2010). A systematic review of evidence for the added benefits to health of exposure to natural environments. *BMC Public Health, 10*(1), 456. https://doi.org/10.1186/1471-2458-10-456

Bowman, A. D., Griffis, J. C., Visscher, K. M., Dobbins, A. C., Gawne, T. J., DiFrancesco, M. W., & Szaflarski, J. P. (2017). Relationship between alpha rhythm and the default mode network: An EEG-fMRI study. *Journal of Clinical Neurophysiology, 34*(6), 527–533. https://doi.org/10.1097/WNP.0000000000000411

Brown, K. W., & Ryan, R. M. (2003). The benefits of being present: Mindfulness and its role in psychological well-being. *Journal of Personality and Social Psychology, 84*(4), 822–848. https://doi.org/10.1037/0022-3514.84.4.822

Buckner, R. L., Andrews-Hanna, J. R., & Schacter, D. L. (2008). The brain's default network: Anatomy, function, and relevance to disease. *Annals of the New York Academy of Sciences, 1124*, 1–38. https://doi.org/10.1196/annals.1440.011

Buckner, R. L., & Vincent, J. L. (2007). Unrest at rest: Default activity and spontaneous network correlations. *NeuroImage, 37*(4), 1091–1096. https://doi.org/10.1016/j.neuroimage.2007.01.010

Christoff, K. (2012). Undirected thought: Neural determinants and correlates. *Brain Research, 1428*, 51–59. https://doi.org/10.1016/j.brainres.2011.09.060

Christoff, K., Gordon, A. M., Smallwood, J., Smith, R., & Schooler, J. W. (2009). Experience sampling during fMRI reveals default network and executive system contributions to mind wandering. *Proceedings of the National Academy of Sciences, 106*(21), 8719–8724. https://doi.org/10.1073/pnas.0900234106

Christoff, K., Irving, Z. C., Fox, K. C., Spreng, R. N., & Andrews-Hanna, J. R. (2016). Mind-wandering as spontaneous thought: A dynamic framework. *Nature Reviews Neuroscience, 17*(11), 718–731. https://doi.org/10.1038/nrn.2016.113

Corbetta, M., & Shulman, G. L. (2002). Control of goal-directed and stimulus-driven attention in the brain. *Nature Reviews Neuroscience, 3*(3), 201–215. https://doi.org/10.1038/nrn755

Creswell, J. D., Way, B. M., Eisenberger, N. I., & Lieberman, M. D. (2007). Neural correlates of dispositional mindfulness during affect labeling. *Psychosomatic Medicine, 69*(6), 560–565. https://doi.org/10.1097/PSY.0b013e3180f6171f

Cullen, M. (2011). Mindfulness-based interventions: An emerging phenomenon. *Mindfulness, 2*(3), 186–193. https://doi.org/10.1007/s12671-011-0058-1

Damoiseaux, J. S., Rombouts, S. A., Barkhof, F., Scheltens, P., Stam, C. J., Smith, S. M., & Beckmann, C. F. (2006). Consistent resting-state networks across healthy subjects. *Proceedings of the National Academy of Sciences, 103*(37), 13848–13853. https://doi.org/10.1073/pnas.0601417103

de Kort, Y. D., Meijnders, A., Sponselee, A., & Ijsselsteijn, W. (2006). What's wrong with virtual trees? Restoring from stress in a mediated environment. *Journal of Environmental Psychology, 26*(4), 309–320. https://doi.org/10.1016/j.jenvp.2006.09.001

De Luca, M., Beckmann, C. F., De Stefano, N., Matthews, P. M., & Smith, S. M. (2006). fMRI resting state networks define distinct modes of long-distance interactions in the human brain. *NeuroImage, 29*(4), 1359–1367. https://doi.org/10.1016/j.neuroimage.2005.08.035

Drews, F. A., Pasupathi, M., & Strayer, D. L. (2008). Passenger and cell phone conversations in simulated driving. *Journal of Experimental Psychology: Applied, 14*(4), 392–400. https://doi.org/10.1037/a0013119

Eberth, J., & Sedlmeier, P. (2012). The effects of mindfulness meditation: A meta-analysis. *Mindfulness, 3*(3), 174–189. https://doi.org/10.1007/s12671-012-0101-x

Escera, C., Alho, K., Winkler, I., & Näätänen, R. (1998). Neural mechanisms of involuntary attention to acoustic novelty and change. *Journal of Cognitive Neuroscience, 10*(5), 590–604. https://doi.org/10.1162/089892998562997

Faber Taylor, A., & Kuo, F. E. M. (2011). Could exposure to everyday green spaces help treat ADHD? Evidence from children's play settings. *Applied Psychology: Health and Well-Being, 3*(3), 281–303. https://doi.org/10.1111/j.1758-0854.2011.01052.x

Felsten, G. (2009). Where to take a study break on the college campus: An attention restoration theory perspective. *Journal of Environmental Psychology, 29*(1), 160–167. https://doi.org/10.1016/j.jenvp.2008.11.006

Förster, J., Friedman, R. S., & Liberman, N. (2004). Temporal construal effects on abstract and concrete thinking: Consequences for insight and creative cognition. *Journal of Personality and Social Psychology, 87*(2), 177–189. https://doi.org/10.1037/0022-3514.87.2.177

Fox, M. D., Snyder, A. Z., Vincent, J. L., Corbetta, M., Van Essen, D. C., & Raichle, M. E. (2005). The human brain is intrinsically organized into dynamic, anti-correlated functional networks. *Proceedings of the National Academy of Sciences, 102*(27), 9673–9678. https://doi.org/10.1073/pnas.0504136102

Fransson, P. (2005). Spontaneous low-frequency BOLD signal fluctuations: An fMRI investigation of the resting-state default mode of brain function hypothesis. *Human Brain Mapping, 26*(1), 15–29. https://doi.org/10.1002/hbm.20113

Fransson, P. (2006). How default is the default mode of brain function? Further evidence from intrinsic BOLD signal fluctuations. *Neuropsychologia, 44*(14), 2836–2845. https://doi.org/10.1016/j.neuropsychologia.2006.06.017

Fransson, P., & Marrelec, G. (2008). The precuneus/posterior cingulate cortex plays a pivotal role in the default mode network: Evidence from a partial correlation network analysis. *NeuroImage, 42*(3), 1178–1184. https://doi.org/10.1016/j.neuroimage.2008.05.059

Fuller, R. A., Irvine, K. N., Devine-Wright, P., Warren, P. H., & Gaston, K. J. (2007). Psychological benefits of greenspace increase with biodiversity. *Biology Letters*, 3(4), 390–394. https://doi.org/10.1098/rsbl.2007.0149

Ghatan, P. H., Hsieh, J. C., Petersson, K. M., Stone-Elander, S., & Ingvar, M. (1998). Coexistence of attention-based facilitation and inhibition in the human cortex. *NeuroImage*, 7(1), 23–29. https://doi.org/10.1006/nimg.1997.0307

Gidlow, C. J., Jones, M. V., Hurst, G., Masterson, D., Clark-Carter, D., Tarvainen, M. P., Smith, G., & Nieuwenhuijsen, M. (2016). Where to put your best foot forward: Psycho-physiological responses to walking in natural and urban environments. *Journal of Environmental Psychology*, 45, 22–29. https://doi.org/10.1016/j.jenvp.2015.11.003

Greenwood, A., & Gatersleben, B. (2016). Let's go outside! Environmental restoration amongst adolescents and the impact of friends and phones. *Journal of Environmental Psychology*, 48, 131–139. https://doi.org/10.1016/j.jenvp.2016.09.007

Greicius, M. D., Krasnow, B., Reiss, A. L., & Menon, V. (2003). Functional connectivity in the resting brain: A network analysis of the default mode hypothesis. *Proceedings of the National Academy of Sciences*, 100(1), 253–258. https://doi.org/10.1073/pnas.0135058100

Hampson, M., Driesen, N. R., Skudlarski, P., Gore, J. C., & Constable, R. T. (2006). Brain connectivity related to working memory performance. *Journal of Neuroscience*, 26(51), 13338–13343. https://doi.org/10.1523/JNEUROSCI.3408-06.2006

Hartig, T., Evans, G. W., Jamner, L. D., Davis, D. S., & Gärling, T. (2003). Tracking restoration in natural and urban field settings. *Journal of Environmental Psychology*, 23(2), 109–123. https://doi.org/10.1016/S0272-4944(02)00109-3

Hartig, T., Kaiser, F. G., & Bowler, P. A. (2001). Psychological restoration in nature as a positive motivation for ecological behavior. *Environment and Behavior*, 33(4), 590–607. https://doi.org/10.1177/00139160121973142

Hasenkamp, W., Wilson-Mendenhall, C. D., Duncan, E., & Barsalou, L. W. (2012). Mind wandering and attention during focused meditation: A fine-grained temporal analysis of fluctuating cognitive states. *NeuroImage*, 59(1), 750–760. https://doi.org/10.1016/j.neuroimage.2011.07.008

Herzog, T. R., Black, A. M., Fountaine, K. A., & Knotts, D. J. (1997). Reflection and attentional recovery as distinctive benefits of restorative environments. *Journal of Environmental Psychology*, 17(2), 165–170. https://doi.org/10.1006/jevp.1997.0051

Herzog, T. R., Herbert, E. J., Kaplan, R., & Crooks, C. L. (2000). Cultural and developmental comparisons of landscape perceptions and preferences. *Environment and Behavior*, 32(3), 323–346. https://doi.org/10.1177/0013916500323002

Herzog, T. R., Maguire, P., & Nebel, M. B. (2003). Assessing the restorative components of environments. *Journal of Environmental Psychology*, 23(2), 159–170. https://doi.org/10.1016/S0272-4944(02)00113-5

Hillman, C. H., Erickson, K. I., & Kramer, A. F. (2008). Be smart, exercise your heart: Exercise effects on brain and cognition. *Nature Reviews Neuroscience*, 9(1), 58–65. https://doi.org/10.1038/nrn2298

Hills, A. P., King, N. A., & Armstrong, T. P. (2007). The contribution of physical activity and sedentary behaviours to the growth and development of children and adolescents. *Sports Medicine*, 37(6), 533–545. https://doi.org/10.2165/00007256-200737060-00006

Hoffman, L. R., Harburg, E., & Maier, N. R. (1962). Differences and disagreement as factors in creative group problem solving. *Journal of Abnormal and Social Psychology, 64*(3), 206–214. https://doi.org/10.1037/h0045952

Interrante, V., Ries, B., & Anderson, L. (2006, March 25–29). Distance perception in immersive virtual environments, revisited. In B. Frölich, D. Bowman, & H. Iwata (Program Chairs), *IEEE Virtual Reality: Conference proceedings* (pp. 3–10). IEEE.

James, W. (1892). *Psychology: The briefer course.* Holt. https://doi.org/10.1037/11060-000

Jang, J. H., Jung, W. H., Kang, D. H., Byun, M. S., Kwon, S. J., Choi, C. H., & Kwon, J. S. (2011). Increased default mode network connectivity associated with meditation. *Neuroscience Letters, 487*(3), 358–362. https://doi.org/10.1016/j.neulet.2010.10.056

Kahn, P. H., Friedman, B., Gill, B., Hagman, J., Severson, R. L., Freier, N. G., Feldman, E. N., Carrèr, S., & Stolyar, A. (2008). A plasma display window? The shifting baseline problem in a technologically mediated natural world. *Journal of Environmental Psychology, 28*(2), 192–199. https://doi.org/10.1016/j.jenvp.2007.10.008

Kandalaft, M. R., Didehbani, N., Krawczyk, D. C., Allen, T. T., & Chapman, S. B. (2013). Virtual reality social cognition training for young adults with high-functioning autism. *Journal of Autism and Developmental Disorders, 43*(1), 34–44. https://doi.org/10.1007/s10803-012-1544-6

Kaplan, R., & Kaplan, S. (2011). Well-being, reasonableness, and the natural environment. *Applied Psychology: Health and Well-Being, 3*(3), 304–321. https://doi.org/10.1111/j.1758-0854.2011.01055.x

Kaplan, R. J., Greenwood, C. E., Winocur, G., & Wolever, T. M. (2000). Cognitive performance is associated with glucose regulation in healthy elderly persons and can be enhanced with glucose and dietary carbohydrates. *American Journal of Clinical Nutrition, 72*(3), 825–836. https://doi.org/10.1093/ajcn/72.3.825

Kaplan, S. (1995). The restorative benefits of nature: Toward an integrative framework. *Journal of Environmental Psychology, 15*(3), 169–182. https://doi.org/10.1016/0272-4944(95)90001-2

Kaplan, S. (2001). Meditation, restoration, and the management of mental fatigue. *Environment and Behavior, 33*(4), 480–506. https://doi.org/10.1177/00139160121973106

Kaplan, S., & Berman, M. G. (2010). Directed attention as a common resource for executive functioning and self-regulation. *Perspectives on Psychological Science, 5*(1), 43–57. https://doi.org/10.1177/1745691609356784

Kennedy, D. O., & Scholey, A. B. (2000). Glucose administration, heart rate and cognitive performance: Effects of increasing mental effort. *Psychopharmacology, 149*(1), 63–71. https://doi.org/10.1007/s002139900335

Korpela, K., De Bloom, J., Sianoja, M., Pasanen, T., & Kinnunen, U. (2017). Nature at home and at work: Naturally good? Links between window views, indoor plants, outdoor activities and employee well-being over one year. *Landscape and Urban Planning, 160*, 38–47. https://doi.org/10.1016/j.landurbplan.2016.12.005

Laufs, H., Krakow, K., Sterzer, P., Eger, E., Beyerle, A., Salek-Haddadi, A., & Kleinschmidt, A. (2003). Electroencephalographic signatures of attentional and cognitive default modes in spontaneous brain activity fluctuations at rest. *Proceedings of the National Academy of Sciences, 100*(19), 11053–11058. https://doi.org/10.1073/pnas.1831638100

Laumann, K., Gärling, T., & Stormark, K. M. (2001). Rating scale measures of restorative components of environments. *Journal of Environmental Psychology, 21*(1), 31–44. https://doi.org/10.1006/jevp.2000.0179

Laumann, K., Gärling, T., & Stormark, K. M. (2003). Selective attention and heart rate responses to natural and urban environments. *Journal of Environmental Psychology, 23*(2), 125–134. https://doi.org/10.1016/S0272-4944(02)00110-X

Li, D., & Sullivan, W. C. (2016). Impact of views to school landscapes on recovery from stress and mental fatigue. *Landscape and Urban Planning, 148*, 149–158. https://doi.org/10.1016/j.landurbplan.2015.12.015

Liang, X., Zou, Q., He, Y., & Yang, Y. (2016). Topologically reorganized connectivity architecture of default-mode, executive-control, and salience networks across working memory task loads. *Cerebral Cortex, 26*(4), 1501–1511. https://doi.org/10.1093/cercor/bhu316

Lutz, J., Herwig, U., Opialla, S., Hittmeyer, A., Jäncke, L., Rufer, M., Holtforth, M. G., & Brühl, A. B. (2014). Mindfulness and emotion regulation—An fMRI study. *Social Cognitive and Affective Neuroscience, 9*(6), 776–785.

Lymeus, F., Lindberg, P., & Hartig, T. (2018). Building mindfulness bottom-up: Meditation in natural settings supports open monitoring and attention restoration. *Consciousness and Cognition, 59*, 40–56. https://doi.org/10.1016/j.concog.2018.01.008

Lymeus, F., Lundgren, T., & Hartig, T. (2017). Attentional effort of beginning mindfulness training is offset with practice directed toward images of natural scenery. *Environment and Behavior, 49*(5), 536–559. https://doi.org/10.1177/0013916516657390

Maier, N. R. F. (1958). *The appraisal interview: Objectives, methods and skills.* Wiley.

Maier, N. R. F., & Solem, A. R. (1962). Improving solutions by turning choice situations into problems. *Personnel Psychology, 15*(2), 151–157. https://doi.org/10.1111/j.1744-6570.1962.tb01857.x

Mason, M. F., Norton, M. I., Van Horn, J. D., Wegner, D. M., Grafton, S. T., & Macrae, C. N. (2007). Wandering minds: The default network and stimulus-independent thought. *Science, 315*(5810), 393–395. https://doi.org/10.1126/science.1131295

Mattay, V. S., Callicott, J. H., Bertolino, A., Heaton, I., Frank, J. A., Coppola, R., Berman, K. F., Goldberg, T. E., & Weinberger, D. R. (2000). Effects of dextroamphetamine on cognitive performance and cortical activation. *NeuroImage, 12*(3), 268–275. https://doi.org/10.1006/nimg.2000.0610

Mazoyer, B., Zago, L., Mellet, E., Bricogne, S., Etard, O., Houdé, O., Crivello, F., Joliot, M., Petit, L., & Tzourio-Mazoyer, N. (2001). Cortical networks for working memory and executive functions sustain the conscious resting state in man. *Brain Research Bulletin, 54*(3), 287–298. https://doi.org/10.1016/S0361-9230(00)00437-8

Mednick, S. (1962). The associative basis of the creative process. *Psychological Review, 69*(3), 220–232. https://doi.org/10.1037/h0048850

Meiklejohn, J., Phillips, C., Freedman, M. L., Griffin, M. L., Biegel, G., Roach, A., Frank, J., Burke, C., Pinger, L., Soloway, G., Isberg, R., Sibinga, E., Grossman, L., & Saltzman, A. (2012). Integrating mindfulness training into K–12 education: Fostering the resilience of teachers and students. *Mindfulness, 3*(4), 291–307. https://doi.org/10.1007/s12671-012-0094-5

Mittner, M., Hawkins, G. E., Boekel, W., & Forstmann, B. U. (2016). A neural model of mind wandering. *Trends in Cognitive Sciences, 20*(8), 570–578. https://doi.org/

10.1016/j.tics.2016.06.004 (Correction published 2017, *Trends in Cognitive Sciences, 21*(6), p. 489)

Nieoullon, A. (2002). Dopamine and the regulation of cognition and attention. *Progress in Neurobiology, 67*(1), 53–83. https://doi.org/10.1016/S0301-0082(02)00011-4

Ohly, H., White, M. P., Wheeler, B. W., Bethel, A., Ukoumunne, O. C., Nikolaou, V., & Garside, R. (2016). Attention restoration theory: A systematic review of the attention restoration potential of exposure to natural environments. *Journal of Toxicology and Environmental Health, Part B: Critical Reviews, 19*(7), 305–343. https://doi.org/10.1080/10937404.2016.1196155

Onton, J., Delorme, A., & Makeig, S. (2005). Frontal midline EEG dynamics during working memory. *NeuroImage, 27*(2), 341–356. https://doi.org/10.1016/j.neuroimage.2005.04.014

Raichle, M. E., & Snyder, A. Z. (2007). A default mode of brain function: A brief history of an evolving idea. *NeuroImage, 37*(4), 1083–1090. https://doi.org/10.1016/j.neuroimage.2007.02.041

Ramel, W., Goldin, P. R., Carmona, P. E., & McQuaid, J. R. (2004). The effects of mindfulness meditation on cognitive processes and affect in patients with past depression. *Cognitive Therapy and Research, 28*(4), 433–455. https://doi.org/10.1023/B:COTR.0000045557.15923.96

Rizzo, A., Hartholt, A., Grimani, M., Leeds, A., & Liewer, M. (2014). Virtual reality exposure therapy for combat-related posttraumatic stress disorder. *Computer, 47*(7), 31–37. https://doi.org/10.1109/MC.2014.199

Robertson, I. H., Manly, T., Andrade, J., Baddeley, B. T., & Yiend, J. (1997). Oops! Performance correlates of everyday attentional failures in traumatic brain injured and normal subjects. *Neuropsychologia, 35*(6), 747–758.

Roe, J. J., Aspinall, P. A., Mavros, P., & Coyne, R. (2013). Engaging the brain: The impact of natural versus urban scenes using novel EEG methods in an experimental setting. *Environmental Sciences, 1*(2), 93–104. https://doi.org/10.12988/es.2013.3109

Roiser, J. P., Müller, U., Clark, L., & Sahakian, B. J. (2007). The effects of acute tryptophan depletion and serotonin transporter polymorphism on emotional processing in memory and attention. *International Journal of Neuropsychopharmacology, 10*(4), 449–461. https://doi.org/10.1017/S146114570600705X

Rosenbaum, M. S., Otalora, M. L., & Ramírez, G. C. (2016). The restorative potential of shopping malls. *Journal of Retailing and Consumer Services, 31*, 157–165. https://doi.org/10.1016/j.jretconser.2016.02.011

Rothbaum, B. O., Hodges, L. F., Kooper, R., Opdyke, D., Williford, J. S., & North, M. (1995). Effectiveness of computer-generated (virtual reality) graded exposure in the treatment of acrophobia. *American Journal of Psychiatry, 152*, 626–628. https://doi.org/10.1176/ajp.152.4.626

Safren, S. A., Otto, M. W., Sprich, S., Winett, C. L., Wilens, T. E., & Biederman, J. (2005). Cognitive-behavioral therapy for ADHD in medication-treated adults with continued symptoms. *Behaviour Research and Therapy, 43*(7), 831–842. https://doi.org/10.1016/j.brat.2004.07.001

Sahlin, E., Matuszczyk, J. V., Ahlborg, G., Jr., & Grahn, P. (2012). How do participants in nature-based therapy experience and evaluate their rehabilitation? *Journal of Therapeutic Horticulture, 22*(1), 8–22.

Sanchez-Vives, M. V., & Slater, M. (2005). From presence to consciousness through virtual reality. *Nature Reviews Neuroscience, 6,* 332–339. https://doi.org/10.1038/nrn1651

Schnall, S., Zadra, J. R., & Proffitt, D. R. (2010). Direct evidence for the economy of action: Glucose and the perception of geographical slant. *Perception, 39*(4), 464–482. https://doi.org/10.1068/p6445

Schooler, J. W., Smallwood, J., Christoff, K., Handy, T. C., Reichle, E. D., & Sayette, M. A. (2011). Meta-awareness, perceptual decoupling and the wandering mind. *Trends in Cognitive Sciences, 15*(7), 319–326. https://doi.org/10.1016/j.tics.2011.05.006

Sherrington, C. S. (1906). Observations on the scratch-reflex in the spinal dog. *Journal of Physiology, 34*(1–2), 1–50. https://doi.org/10.1113/jphysiol.1906.sp001139

Sio, U. N., & Ormerod, T. C. (2009). Does incubation enhance problem solving? A meta-analytic review. *Psychological Bulletin, 135*(1), 94–120. https://doi.org/10.1037/a0014212

Smallwood, J., Brown, K., Baird, B., & Schooler, J. W. (2012). Cooperation between the default mode network and the frontal-parietal network in the production of an internal train of thought. *Brain Research, 1428,* 60–70. https://doi.org/10.1016/j.brainres.2011.03.072

Smallwood, J., McSpadden, M., & Schooler, J. W. (2008). When attention matters: The curious incident of the wandering mind. *Memory & Cognition, 36*(6), 1144–1150. https://doi.org/10.3758/MC.36.6.1144

Spendrup, S., Hunter, E., & Isgren, E. (2016). Exploring the relationship between nature sounds, connectedness to nature, mood and willingness to buy sustainable food: A retail field experiment. *Appetite, 100,* 133–141. https://doi.org/10.1016/j.appet.2016.02.007

Stigsdotter, U. K., Corazon, S. S., Sidenius, U., Refshauge, A. D., & Grahn, P. (2017). Forest design for mental health promotion—Using perceived sensory dimensions to elicit restorative responses. *Landscape and Urban Planning, 160,* 1–15. https://doi.org/10.1016/j.landurbplan.2016.11.012

Strayer, D. L., & Drews, F. A. (2007). Cell-phone–induced driver distraction. *Current Directions in Psychological Science, 16*(3), 128–131. https://doi.org/10.1111/j.1467-8721.2007.00489.x

Strayer, D. L., Drews, F. A., & Johnston, W. A. (2003). Cell phone-induced failures of visual attention during simulated driving. *Journal of Experimental Psychology: Applied, 9*(1), 23–32. https://doi.org/10.1037/1076-898X.9.1.23

Teipel, S. J., Bokde, A. L., Meindl, T., Amaro, E., Jr., Soldner, J., Reiser, M. F., Herpertz, S. C., Möller, H. J., & Hampel, H. (2010). White matter microstructure underlying default mode network connectivity in the human brain. *NeuroImage, 49*(3), 2021–2032. https://doi.org/10.1016/j.neuroimage.2009.10.067

Tennessen, C. M., & Cimprich, B. (1995). Views to nature: Effects on nature. *Journal of Environmental Psychology, 15,* 77–85. https://doi.org/10.1016/0272-4944(95)90016-0

Turner, M. L., & Engle, R. W. (1989). Is working memory capacity task dependent? *Journal of Memory & Language, 28,* 127–154. https://doi.org/10.1016/0749-596X(89)90040-5

Twedt, E., Rainey, R. M., & Proffitt, D. R. (2016). Designed natural spaces: Informal gardens are perceived to be more restorative than formal gardens. *Frontiers in Psychology, 7*, 88. https://doi.org/10.3389/fpsyg.2016.00088

Uddin, L. Q., Kelly, A. M., Biswal, B. B., Castellanos, F. X., & Milham, M. P. (2009). Functional connectivity of default mode network components: Correlation, anticorrelation, and causality. *Human Brain Mapping, 30*(2), 625–637. https://doi.org/10.1002/hbm.20531

Uhls, Y. T., Michikyan, M., Morris, J., Garcia, D., Small, G. W., Zgourou, E., & Greenfield, P. M. (2014). Five days at outdoor education camp without screens improves preteen skills with nonverbal emotion cues. *Computers in Human Behavior, 39*, 387–392. https://doi.org/10.1016/j.chb.2014.05.036

Ulrich, R. S. (1983). Aesthetic and affective response to natural environment. In I. Altman & J. R. Wohlwill (Eds.), *Behavior and the natural environment* (Vol. 6, pp. 85–125). Springer. https://doi.org/10.1007/978-1-4613-3539-9_4

Ulrich, R. S., Simons, R. F., Losito, B. D., Fiorito, E., Miles, M. A., & Zelson, M. (1991). Stress recovery during exposure to natural and urban environments. *Journal of Environmental Psychology, 11*(3), 201–230. https://doi.org/10.1016/S0272-4944(05)80184-7

Valtchanov, D., Barton, K. R., & Ellard, C. (2010). Restorative effects of virtual nature settings. *Cyberpsychology, Behavior, and Social Networking, 13*(5), 503–512. https://doi.org/10.1089/cyber.2009.0308

Van Cauter, E., Blackman, J. D., Roland, D., Spire, J. P., Refetoff, S., & Polonsky, K. S. (1991). Modulation of glucose regulation and insulin secretion by circadian rhythmicity and sleep. *Journal of Clinical Investigation, 88*(3), 934–942. https://doi.org/10.1172/JCI115396

Van Cauter, E., Polonsky, K. S., & Scheen, A. J. (1997). Roles of circadian rhythmicity and sleep in human glucose regulation. *Endocrine Reviews, 18*(5), 716–738.

Van den Berg, A. E., Jorgensen, A., & Wilson, E. R. (2014). Evaluating restoration in urban green spaces: Does setting type make a difference? *Landscape and Urban Planning, 127*, 173–181. https://doi.org/10.1016/j.landurbplan.2014.04.012

van den Heuvel, M. P., & Hulshoff Pol, H. E. (2010). Exploring the brain network: A review on resting-state fMRI functional connectivity. *European Neuropsychopharmacology, 20*(8), 519–534. https://doi.org/10.1016/j.euroneuro.2010.03.008

van Rompay, T. J., & Jol, T. (2016). Wild and free: Unpredictability and spaciousness as predictors of creative performance. *Journal of Environmental Psychology, 48*, 140–148. https://doi.org/10.1016/j.jenvp.2016.10.001

Vincent, J. L., Patel, G. H., Fox, M. D., Snyder, A. Z., Baker, J. T., Van Essen, D. C., Zempel, J. M., Snyder, L. H., Corbetta, M., & Raichle, M. E. (2007). Intrinsic functional architecture in the anaesthetized monkey brain. *Nature, 447*(7140), 83–86. https://doi.org/10.1038/nature05758

Wagenfeld, A., Roy-Fisher, C., & Mitchell, C. (2013). Collaborative design: Outdoor environments for veterans with PTSD. *Facilities, 31*(9/10), 391–406. https://doi.org/10.1108/02632771311324954

Wascher, E., Rasch, B., Sänger, J., Hoffmann, S., Schneider, D., Rinkenauer, G., Heuer, H., & Gutberlet, I. (2014). Frontal theta activity reflects distinct aspects of mental fatigue. *Biological Psychology, 96*, 57–65. https://doi.org/10.1016/j.biopsycho.2013.11.010

Wechsler, D. (1987). *Wechsler Memory Scale manual.* Psychological Corporation.

Wesnes, K. A., Pincock, C., Richardson, D., Helm, G., & Hails, S. (2003). Breakfast reduces declines in attention and memory over the morning in schoolchildren. *Appetite, 41*(3), 329–331. https://doi.org/10.1016/j.appet.2003.08.009

Willemsen, P., & Gooch, A. A. (2002, March 24–28). Perceived egocentric distances in real, image-based, and traditional virtual environments. In B. Loftin, J. X. Chen, S. Rizzo, M. Goebel, & M. Hirose (Eds.), *IEEE Virtual Reality 2002: Proceedings* (pp. 275–276). IEEE.

Wilson, E. O. (1984). *Biophilia.* Harvard University Press.

Zedelius, C. M., & Schooler, J. W. (2015). Mind wandering "ahas" versus mindful reasoning: Alternative routes to creative solutions. *Frontiers in Psychology, 6,* 834. https://doi.org/10.3389/fpsyg.2015.00834

Zhang, J. W., Piff, P. K., Iyer, R., Koleva, S., & Keltner, D. (2014). An occasion for unselfing: Beautiful nature leads to prosociality. *Journal of Environmental Psychology, 37,* 61–72. https://doi.org/10.1016/j.jenvp.2013.11.008

Zickuhr, K., & Smith, A. (2012, April 13). *Digital differences.* Pew Research Center. https://www.pewresearch.org/internet/2012/04/13/digital-differences/

8

CHARTING A WAY FORWARD

Navigating the Attention Economy

SEAN LANE, PAUL ATCHLEY, AND KACIE MENNIE

What information consumes is rather obvious: it consumes the attention of its recipients. Hence a wealth of information creates a poverty of attention and a need to allocate that attention efficiently among the overabundance of information sources that might consume it.

—Herbert Simon (1971)

The previous chapters in this book have provided a rich and provocative view of research exploring how information technology (IT) influences the way that we think, feel, and behave. As made clear in our Introduction and in Simon's prescient comments, this impact comes about because IT is purposely designed to attract and hold attention. Technology focuses our attention on some aspects of our experience and away from others. So, for instance, we might focus on a phone conversation rather than an obstacle in the road (see Chapter 6), on a YouTube video rather than the facial expressions and body language of our friends (see Chapter 3), or on the process of taking a photograph rather than the experience of viewing art (see Chapter 5).

https://doi.org/10.1037/0000208-009
Human Capacity in the Attention Economy, S. Lane and P. Atchley (Editors)

In this final chapter, we take a step back to consider how we can better navigate the complex landscape of the attention economy. We begin by briefly considering the near future, in which advances in artificial intelligence (AI) and robotics promise even stronger technological influences on our everyday lives. The challenges identified in the first main section of this chapter include a number of issues already considered in previous chapters and other ones, such as the impact of technology on our sense of identity and security. In the section "Changing the Likelihood That Technology Will Positively Impact Our Lives," we ask what can be done to increase the likelihood that IT enhances our lives rather than detracts from it. We take the position that it is possible to tip the balance in our favor when people assume a more active role in defining their relationship to technology. More broadly, achieving such a goal will need to be made through efforts that range from the individual to organizational, societal and cultural. In the final section, we summarize our thoughts about the way forward, including the need for continued research to provide a sound basis for the decisions that lie ahead.

WHAT ARE THE THREATS AND OPPORTUNITIES THAT COME FROM ADVANCING TECHNOLOGY?

The focus of this volume has been on human cognitive, emotional, and behavioral capacities, and the extent to which these capacities have been challenged by rapid changes in IT over the past 25 years. But what will the future bring? The answer to that general question depends on whom you ask, the specific form and focus of the question, and the time horizon for prediction. For the purposes of this chapter, we focus on the types of issues that are anticipated in the next 10 to 20 years due to new developments in AI and robotics because there is much less consensus about the challenges that will occur over longer periods.

Broad Impacts of an Increasingly Digital Life

The Pew Research Center (J. Anderson & Rainie, 2018) recently asked a group of more than 1,100 experts a single question about how changes in digital life in the next 10 years will impact people's well-being. The majority of respondents indicated that more good than harm will come out of these changes. However, themes emerged from an evaluation of the written responses that characterized specific aspects of life that they believed would likely be enhanced versus harmed by technological advances. The areas in which the experts saw the potential for expected improvements included

- greater human connection and access to people and information across geographic boundaries;
- more effective interactions between people and organizations (government and business);
- knowledge and resources intelligently delivered just in time;
- a stronger ability for people to change and improve their lives and pursue positive, meaningful goals; and
- greater usefulness and efficacy of digital tools that integrate more closely with human life.

For instance, some experts, for example, Daniel Weitzner, have argued that improved connections between people would be a key driver of improvements to well-being:

> Whether on questions of politics, community affairs, science, education, romance or economic life, the internet does connect people with meaningful and rewarding information and relationships. . . . I have to feel confident that we can continue to gain fulfillment from these human connections. (J. Anderson & Rainie, 2018, para. 9)

Stephen Downes, among others, expressed the view that technology would fuel the drive toward constructive, life-affirming goals rather than less constructive ones:

> The internet will help rather than harm people's well-being because it breaks down barriers and supports them in their ambitions and objectives. We see a lot of disruption today caused by this feature, as individuals and companies act out a number of their less desirable ambitions and objectives. Racism, intolerance, greed and criminality have always lurked beneath the surface, and it is no surprise to see them surface. But the vast majority of human ambitions and objectives are far more noble: people desire to educate themselves, people desire to communicate with others, people desire to share their experiences, people desire to create networks of enterprise, commerce and culture. . . . (J. Anderson & Rainie, 2018, para. 12)

However, experts' responses were not uniformly positive about the future. Respondents expressed a number of concerns about the potential harmful impacts on everyday life, including

- negative changes to our cognitive abilities, such as memory, creativity, and analytical thinking as a result of overuse and overreliance on technology;

- addiction to technological tools designed to take advantage of human reward systems;

- decreased agency or control and a sense of distrust caused by systems that have the ability to monitor and shape our behavior;

- emotional distress resulting from an overabundance of information, poor usability, and reduced physical interaction between people; and

- threats to society and individuals that are made more serious by increased digital connectedness and advancing technology (e.g., cybercrime, biased algorithms, AI's impact on jobs).

For example, experts such as Jason Hong anticipate that problems created by current technology—designed with commercial goals in mind—are likely to continue:

> Today, we now also have organizations that are actively vying for our attention, distracting us with smartphone notifications, highly personalized news, addictive games, Buzzfeed-style headlines and fake news. These organizations also have a strong incentive to optimize their interaction loops, drawing on techniques from psychology and mass A/B testing [a randomized experiment comparing two versions of a website or application] to draw us in. Most of the time it's to increase click-through rates, daily active users and other engagement metrics, and ultimately to increase revenues. There are two major problems with these kinds of interactions. The first is just feeling stressed all the time, due to a constant stream of interruptions combined with fear of missing out. The second, and far more important, is that engagement with this kind of content means that we are spending less time building and maintaining relationships with actual people. (J. Anderson & Rainie, 2018, para. 17)

Other experts such as Judith Donath pointed out that improvements in the ability of computers to interact with human beings (e.g., from an increased breadth of sensors) also leads to an increased ability to observe and influence our thoughts, emotions, and behaviors:

> We will see a big increase in the ability of technologies to affect our sense of well-being. The ability to both monitor and manipulate individuals is rapidly increasing. Over the past decade, technologies to track our online behavior were perfected; the next decade will see massively increased surveillance of our off-line behavior. It's already commonplace for our physical location, heart rate, etc., to be tracked; voice input provides data not only about what we're saying, but also the affective component of our speech; virtual assistants learn our household habits. The combination of these technologies makes it possible for observers (Amazon, government, Facebook, etc.) to know what we are doing, what is happening around us, and how we react to it all. . . . (J. Anderson & Rainie, 2018, para. 16)

Altogether, responses from the Pew Research Center (J. Anderson & Rainie, 2018) survey paint a picture of a not-too-distant future in which the promise of technology is realized through better design of technology and human systems, and stronger integration between the two. However, they also highlight a number of concerns raised in earlier chapters in this volume, such as addiction (see Chapter 2), emotional distress (see Chapter 3), and changes to cognitive performance (see Chapters 4, 5, 6, and 7). They also describe other, broader threats to people's sense of identity, autonomy, and privacy. Next, we next describe one major threat in this domain, namely, the potential impact of automation on human work.

The Influence of AI and Robotics on Human Jobs

On November 8, 2018, it was announced that the world's first AI news anchor, modeled on a human news reader and aided by machine learning, would begin working for China's Xinhua News Agency (Handley, 2018). Announcements of AI "replacements" have become commonplace in the news media and have spawned concern about the future of a whole range of jobs, including those usually considered high skill, such as a news anchor. However, just how many jobs will be affected, and what types, has been hotly debated. In a highly cited study, Frey and Osborne (2017; originally disseminated in 2013) evaluated the nature of 702 occupational categories and estimated that 47% of U.S. employees work in occupations that are in the high-risk category for disruption by automation likely within the next 20 years. Occupations requiring social and emotional intelligence (e.g., management), and those requiring the generation of novel and creative ideas or products (e.g., scientists, industrial designers), are among those with the lowest risk. In addition, the likelihood of automation decreases with the level of education required for the position and the wages currently paid to individuals in that occupation.

Subsequent analyses generally provide lower estimates of likely job loss. For example, a later study (Arntz et al., 2016) used the results of the Survey of Adult Skills (Organisation for Economic Co-operation and Development [OECD], 2013) to develop a more fine-grained evaluation of occupations, resulting in an estimate of approximately 9% U.S. job loss due to automation. In addition, research using data from 32 OECD countries (Nedelkoska & Quintini, 2018) derived an average estimate of job loss of 14%, although there was large between-country variability. Findings indicated job loss will most likely be in manufacturing and agriculture, with other areas at risk, including transportation and food service. As in previous research, automation-related job loss is less likely the greater the education level currently required for the position. A recent McKinsey report (Manyika et al., 2017) evaluating multiple automation scenarios provided a similar estimate for likely job loss (~15%) but also suggested that up to 60% of occupations will see substantial automation of specific work tasks and that as many as 375 million workers worldwide will need to change their occupation by 2030.

Across the literature, several major themes emerge about the impact of automation in the workplace. First, although disagreements about the exact percentage of job loss remain, there is agreement that these losses will be substantial over the next 20 years. Second, although outright job loss is one outcome, the day-to-day activities of most professions will change, often dramatically. Third, improvements in AI and robotics are likely to lead to entirely new career paths. These and other issues have motivated organizations and governments to consider the challenges ahead. For instance, the Pew Research Center recently surveyed a group of experts to get their views on how training programs might help people learn the skills needed to be successful in the workplace of the future (J. Anderson & Rainie, 2017). Although the general consensus was that new systems would evolve to meet these needs, many experts also surmised that current educational and training systems would be unable to provide the needed skills within the next 10 years. Furthermore, automation could potentially create more fundamental threats: job losses that far exceed job gains and a lack of societal and economic structures to deal with the subsequent changes in how people use their time.

The advent of greater automation in the workplace is likely to be disruptive. However, such disruption is more likely to be beneficial if researchers and designers ask more fine-grained questions, such as: How should the work of machines and humans be integrated in the context of a task (see Chapter 6 and Atchley & Lane, 2014, for discussion)? What types of tasks are best allocated to humans versus machines (not simply whether it is possible to automate a task)? Principled answers to these questions are likely to lead to better informed decisions about the degree of automation that is optimal for a particular job, how technology should be designed to enhance human performance, the nature of training, and many other issues.

CHANGING THE LIKELIHOOD THAT TECHNOLOGY WILL POSITIVELY IMPACT OUR LIVES

The foregoing summary highlights only a few of the ways in which human life will be impacted in the coming years by the technological advances that are becoming increasingly central to our daily activities. Although predictions about the outcome of these advances range from wildly optimistic (e.g., Kurzweil, 2006) to largely negative (e.g., Carr, 2010; Turkle, 2011), we believe that much can be done to increase the likelihood that, on balance, such technology will enhance rather than detract from our lives. Furthermore, our opinion is that this desired outcome is most likely to come from a systematic, comprehensive approach that includes

- changing how human beings engage with technology (e.g., mindful decisions about when to use it or educational programs designed to teach relevant skills);

- improving technology to better support human beings (e.g., by designing applications that support human capacities and the pursuit of individual and prosocial goals); and

- societal, political, and economic changes that support human welfare (e.g., laws that provide greater control over personal data or cultural norms about appropriate attention to technology versus human beings).

These three broad themes are consistent with other recent discussions. For example, related topics were identified from the Pew Research Center survey on technology and human well-being (J. Anderson & Rainie, 2018) discussed earlier in the chapter. In their written answers to survey questions, respondents often suggested potential ways to remedy the harmful effects of advancing technology. The researchers characterized suggestions as falling within the following five general categories:

- Redesign societal institutions and processes to ensure that technology serves the needs of human beings.
- Increase human-centered design of technology.
- Create laws and regulations that balance protection of public good and individual needs with the ability to innovate.
- Educate people about technology and its effective use (media literacy).
- Adjust expectations about what it means to be human as technology advances.

However, not all survey respondents were so sanguine. A sixth category of responses was characterized by general pessimism that any intervention would likely lead to a positive outcome for human well-being.

Similar themes are found in the work of an organization called Time Well Spent (http://www.humanetech.com), founded by Tristan Harris, James Williams, and other former IT leaders (e.g., Harris was formerly design ethicist at Google) whose experience has given them intimate knowledge of the ways in which the current IT environment benefits companies rather than users or society. They suggest that four "levers" are needed to change the balance (Center for Humane Technology, n.d.-b):

- Inspire humane design of products.
- Apply political pressure to change corporate behavior.
- Create a cultural awakening to empower consumers.
- Engage employees of technology companies in the effort.

As can be seen, although the details may vary, there is general agreement about the broad strategies needed to positively influence the impact of IT on human life. In the next sections, we discuss specific ways in which people can change how they engage with technology, that technology can be designed to better support human capacities, and that institutions or group norms can create an environment that benefits human welfare over other interests. Given the focus of this chapter, we have chosen these examples to be illustrative rather than provide an exhaustive description of approaches to the topic. We begin by examining how people can begin to exert greater control over the focus of their attention through the choices they make.

Changing How We Engage With Technology

It is possible to change the impact of technology on everyday life by modifying the environment (including software interfaces), disengaging from technology intermittently and thoughtfully, and by learning new skills that allow people to use technology more effectively. The common thread among these methods is that they involve more active and potentially more mindful ways of using technology.

An example of environmental modification comes from people's attempts to cope with the attentional demands of smartphone use. As revealed in a recent survey of smartphone users by Deloitte (2017), the use of these devices is a strongly ingrained habit: An overwhelming percentage of respondents said they check their phone first thing in the morning (89%) and within an hour of going to bed (81%). Yet, users are not uniformly passive in the face of these demands. In this same survey, almost half (47%) of respondents said they took steps to try to reduce their smartphone usage. The strategies they used were primarily designed to avoid cues to action (temptation), such as keeping their smartphones out of sight when meeting with other people (38%), turning off auditory notifications (32%), and turning off their phone at night (26%). Other popular strategies commonly shared on the internet modify the mobile phone interface in ways that reduce distraction and temptation. These strategies range from fairly simple, such as making your screen grayscale instead of color to reduce its salience (e.g., Allan, 2016), to more involved procedures for deleting and hiding applications to create a minimal interface (e.g., Fortin, 2017). These types of defenses against interruption are consistent with the broader psychological literature on self-control, which finds that people who are the most successful at avoiding temptation are those who limit their exposure to cues to action (e.g., Ent et al., 2015).

Another strategy is to disengage from IT for longer periods. The National Day of Unplugging (https://www.nationaldayofunplugging.com) is one such event, but others recommend disconnecting more regularly, such as a weekly "technology *Shabbat*" (also called "technology Sabbath"; Shlain, 2013) that involves turning off devices from sunset on Friday until sunset Saturday. The goal of such temporary disengagement is often to connect with others in an environment that is free from normal day-to-day distractions as well as develop a healthier relationship with technology. Still longer periods of disengagement are encouraged in "digital detox vacations," which differ in type. Some involve travel to remote areas that lack cell service (e.g., Bunch, 2018); others provide education about healthy and mindful uses of technology (e.g., http://digitaldetox.org/). In both types, the focus often is on having participants spend time in nature with the consequent benefits to attention and well-being (for a discussion, see Schilhab et al., 2018, and Chapter 7, this volume).

Another type of strategy focuses on building skills to help people use IT more effectively, namely, *digital literacy* (e.g., Alexander et al., 2017). Although the definition and nature of digital literacy curricula vary widely, there are several common elements. The goal of digital literacy programs involves helping students learn to use digitals tools creatively, collaboratively, reflectively, and responsibly (e.g., Alexander et al., 2016). In addition to learning cognitive skills, such as the ability to discriminate between reliable and unreliable sources of information, an important aspect of training involves understanding and practicing *digital citizenship*, the responsible use of technology for oneself, others, and society (e.g., http://www.digitalcitizenship.net/; Common Sense Education, n.d.). Training typically includes norms for conduct (e.g., inappropriate communication like online bullying), legal rights and responsibilities (e.g., the notion of copyright), digital security (e.g., protection from viruses and hacking), and physical and psychological health (e.g., how to avoid harmful patterns of technology use). Thus, participants in such programs learn to think about their use of technology within a broader societal context as well as acquire more practical knowledge about how to behave adaptively in a technological environment that can be expected to continue to evolve.

Changing Technology to Better Support Human Capabilities and Goals

IT has been designed to be extremely effective at taking advantage of human foibles (e.g., see Atchley & Lane, 2014, for a discussion). One group credited with increasing the effectiveness of IT in influencing attention and behavior is the *persuasive technology movement,* led by B. J. Fogg, a psychologist at Stanford University. The Fogg behavior model (e.g., Fogg, 2009) posits that behavior is influenced by the interaction of motivation, ability, and triggers. Although Fogg has argued for the ethical use of these principles in product design (e.g., Fogg, 2002), a number of entrepreneurs and designers who took courses from him and his colleagues in the mid-2000s subsequently created commercial applications for Facebook and other companies that used these principles to increase their ability to attract and hold attention (Stolzoff, 2018). More recently, a backlash against such use of behavioral techniques has taken a variety of forms, such as a 2018 letter to the American Psychological Association from the Children's Screen Time Action Network raising the issue of unethical behavior on the part of psychologists involved in the creation of applications aimed at taking advantage of the vulnerabilities of children (C. A. Anderson et al., 2018) or the formation of groups like the Center for Humane Technology (https://humanetech.com/) that focus on reducing the impact of such techniques and changing the extent to which they are used by corporations.

Although IT can negatively influence cognition, emotion, and behavior, this need not be the case. For example, designers have also created tools to help people reduce technological distraction. In September 2018, in response to public pressure, Apple introduced the application Screen Time for iOS to give users (particularly parents) a better tool to help manage technology use (e.g., Mickle, 2018). Screen Time allows users to get a weekly report of their app usage compared with the previous week as well as set time limits on usage. A similar application is available on the Android mobile operating system. Although the effectiveness of such applications to substantially reduce screen time has not been rigorously evaluated (but see Mark et al., 2018), this has not deterred a substantial number of companies from developing and marketing software to reduce distraction. For example, RescueTime (Android, iOS, Mac OS, Windows) is a task management software package that not only records application use on the phone or desktop computer but also allows users to set and track goals, and block distracting applications or websites. Less extensive browser-based applications, such as

Freedom (iOS, Chrome, Firefox, Opera) and LeechBlock (Chrome, Firefox), provide tracking and website blocking functions. Still other applications are designed to reduce the salience of irrelevant features to reduce distraction through the use of a whole-screen minimal interface. Examples of this approach include the word processing application iA Writer (iOS, Mac, Android, Windows) or the to-do list application MinimaList (iOS).

Some of the same design principles used to ensure that users spend more time on social media sites can also be used to positively impact people's health and well-being. For instance, mobile applications such as Zombies, Run! and Superhero Workout gamify physical fitness as a means of increasing participation. Similarly, the company Boundless Mind was founded by two neuroscientists on the proposition that persuasive technology techniques can be harnessed for companies and nonprofits that are pursuing prosocial goals (e.g., applications to reduce pain; Edwards, 2018). Others have argued that designers need to take a broader perspective when creating new applications and consider whether products facilitate human capabilities. For example, the Center for Humane Technology (n.d.-a) recently created a *Humane Design Guide*, which asks designers to consider whether key human sensitivities—emotional, attention, sensemaking, decision making, social reasoning, and group dynamics—are supported or inhibited by a current technology, as a means of considering options for new, improved designs. Such an approach adds a moral and ethical dimension to design that has been historically undervalued in commercial application development.

Although there has been progress designing technology that better supports human capabilities, designers can go much further. For instance, the fields of human factors, human-centered computing, and user experience design have typically focused on ways to make the user's task easier and more efficient, and have had a number of successes in this realm. Certainly, these fields can continue to contribute to the development of technology that works with, instead of against, human perceptual, cognitive, emotional, and behavioral constraints. Along these same lines, there would be considerable value in increased collaboration between basic scientists and application developers to explore the implications of new findings in fields such as psychology, neuroscience, and learning science. But we believe such progress would be incomplete without a broader consideration of human values when creating new products. The design process must consider the balance of benefits to the user and the creator, and how those benefits are elicited, among other issues. As we move toward an even greater reliance on IT, the impact on our lives will depend on wise, ethical design decisions. Thus, the development of such reasoning skills needs to be emphasized in the education of researchers and designers.

Changing Societal, Political, and Economic Structures

As IT has become more ubiquitous, excitement about its utility has often been mixed with concern. Furthermore, as discussed in previous sections, these concerns are amplified by a future that promises an even greater reliance on technology in our everyday lives. As a tool, IT has the potential to be used in ways that benefit or detract from human freedom, opportunity, and well-being. For a number of issues, it is likely that larger scale legal, political, and economic strategies will be needed to achieve the goal of ensuring that IT has greater benefits to human life. For the purposes of this chapter, we briefly discuss the issues of digital privacy, the increased use of computer algorithms in decision making, and the impact of AI and automation on human work.

Digital Privacy

The ability to selectively choose what information is revealed about you to others and to decide how that information is used is challenged in an environment in which it is possible to track one's geographic location at all times, comments made online are potentially discoverable indefinitely, and the ability to collect data outstrips the ability to maintain data security. Adding to the complexity, the value placed on such privacy differs substantially by country (e.g., Bellman et al., 2004). For instance, European countries have generally passed more restrictive data privacy laws than those in the United States. However, even in the United States, concerns about privacy have been recently heightened by prominent news stories about the release of confidential information either via cyberattack (Whittaker, 2018) or through the inappropriate use of data by third parties, such as the revelation that Cambridge Analytics gained access to the data of 50 million Facebook users in the context of the 2016 presidential election campaign (Granville, 2018). Thus, there have been increasing calls for legislation that provides greater safeguards to users. To date, most new legislation directed at addressing this issue in the United States has been made at the state level; recent examples include California (California Consumer Privacy Act, 2018) and Utah (Electronic Information or Data Privacy Act, 2019).

Other countries have moved more aggressively, the most prominent being the European Parliament, which passed the General Data Protection Regulation (GDPR; European Commission, n.d.) in 2016; enforcement began in May 2018. The data protections in the new law include the requirement that organizations provide simplified privacy policy language, that users provide clear consent before data can be used, that one has the right to obtain personal data from the organization, that one has the "right to be forgotten" (i.e., the ability to remove certain types of internet-accessible data), and that allows for the creation of a stronger enforcement framework for violations. Although the GDPR expressly protects the digital privacy rights of residents of the European Union (EU), it also applies to companies outside the EU if they provide goods and services to EU citizens. For this reason, the impact of the GDPR has already been felt more widely because users from a wide variety of countries received requests from companies and other organizations to provide consent as the deadline neared in 2018. However, the long-term consequences, positive and negative, of broad data privacy legislation such as the GDPR are only beginning to be seen, and it remains unclear at this time whether similar types of laws will be passed in the United States.

Algorithmic Decision Making

The use of machine learning to identify patterns in data and to generate predictions has led to an explosion of applications that range from the movie recommendations to decisions about creditworthiness. Such algorithms have a number of advantages over human decision making, including speed, consistency, and the ability to weight a large number of factors simultaneously. Despite these advantages, these algorithms can also be biased (e.g., Osoba & Welser, 2017). For example, ProPublica (Angwin et al., 2016) examined the outcomes of a criminal risk assessment tool called COMPAS (Correctional Offender Management Profiling for Alternative Sanctions) that is used in a number of jurisdictions to make predictions about recidivism and thus to make decisions about bail amounts, prison sentences, and probation. Their evaluation of Broward County, Florida, data revealed that the algorithm identified Black defendants as more likely to reoffend than White defendants, and White defendants were more often labeled as low risk than Black ones (Northpointe, the company that created the algorithm, has disputed this analysis; see Angwin et al., 2016). Other prominent examples of reported algorithmic bias include Facebook job advertisements that were shown only to people under 40 years of age (in violation of federal law; Angwin, Scheiber, & Tobin, 2017); false identification by Amazon's Rekognition software of 28 members of Congress, primarily people of color, as individuals who had been arrested for a crime (Singer, 2018); and higher car insurance premiums for drivers residing in primarily minority geographic areas than drivers with similar risks who lived in predominately White areas (e.g., Angwin, Larson, et al., 2017). We note that many biases already exist in society, but their existence in algorithms means that such biases are propagated even more systematically.

Biases can be introduced into algorithms in a number of ways (e.g., see Osoba & Welser, 2017). Machine learning techniques depend heavily on the data used to train the software. For instance, if the data are biased (e.g., as a function of biased human decision making), they are likely to be represented by the algorithm. In addition, algorithmic accuracy for subpopulations can be affected by low base rates in the population at large (e.g., facial recognition of minorities). The effectiveness of algorithms' predictions can also degrade if conditions change substantially from training—essentially a stability bias (Baer & Kamalnath, 2017). It is also important to recognize that the appropriate criteria for judging an algorithm is not always clear in many situations (e.g., see different views of fairness that could applied to the impact of COMPAS; Corbett-Davies et al., 2016).

A variety of solutions have been proposed to combat potential bias. Most attempt to address the problem that algorithms are often a black box because of their high complexity, proprietary nature, or both. For example, one way of increasing transparency includes developing algorithms that provide information about the causal reasoning involved in a decision (e.g., Athey, 2015). It has also been proposed that algorithms be regularly audited in ways similar to corporate financial statements (Guszcza et al., 2018). Furthermore, systems that make particularly consequential decisions could include a human as part of the process (Osoba & Welser, 2017). Progress on the legislative front has been slow to date and less than comprehensive. For instance, the GDPR includes a section that addresses bias in automated decision making and provides a (nonbinding) right to an explanation about automated decisions. The United States has no broad laws governing the use of algorithms, but New York City passed a bill in 2017 creating a task force to evaluate recommendations for increasing transparency and accountability of algorithms used by city government (Press Office, City of New York, 2018). More recently, legislators in Washington State filed bills (H.B. 1655, S.B. 5527) to provide state government guidelines for evaluating and purchasing automated decision-making systems (Pangburn, 2019). Altogether, although there is now greater public recognition of the possible issues involved in algorithmic decision making, progress on how to address potential bias has been slow, and the specific nature of solutions is only beginning to be worked out.

AI, Automation, and the World of Work

As discussed earlier in the chapter, advances in AI and robotics are predicted to disrupt work as we currently know it, including eliminating some types of jobs and creating new ones (e.g., Manyika et al., 2017). Although a multitude of issues are likely to arise from these changes, such as how people will support themselves if less work is required of them (e.g., discussions about the use of universal income; see Darrow, 2017), we briefly discuss two issues about the role of training and education: the need to acquire new skills to "partner" more effectively with AI and the changing nature of higher education.

Although researchers have argued that more fine-grained data are needed to be able to adequately predict the impact of automation on specific occupations (e.g., Frank et al., 2019), evidence suggests that skills that allow the collaboration of humans and AI are increasingly important to present-day occupations, and the need for them is likely to grow in the future (e.g., Wilson & Daugherty, 2018). For instance, humans can train software agents (e.g., helping chatbots to interact more effectively with humans), effectively describe how decisions were made by an algorithm (e.g., how certain factors were weighted), and actively ensure that AI systems operate in safe and responsible ways (e.g., anticipating and remedying potential harm from a medical system). Furthermore, skills necessary for successfully using AI become important, such as knowing how best to incorporate input into decision making (e.g., using AI recommendations to make a medical diagnosis) or the ability to use the system as an extension of one's skills (e.g., robotic assisted surgery or AI-assisted piloting). Altogether, these skills are quite different from those required by a traditional machine operator (e.g., a magnetic resonance imaging technician) for a number of reasons, including the ability of the system and the human to learn and adapt in tandem. Finding effective ways of imparting these dynamic skills will be an important goal of researchers and educators.

Higher education in the United States is facing questions about its expense, its effectiveness in preparing students for the workplace, whether it serves to increase social mobility, and many other issues (e.g., Davidson, 2017; Selingo, 2013). Advances in AI and automation are also leading to questions about how to best prepare students for a technological future. One response has been an increase in the number of college graduates in STEM (science, technology, engineering, and mathematics) fields in the past decade and fewer graduates in a number of other fields, including the humanities and education (e.g., Nietzel, 2019). However, others (e.g., Anders, 2017; Aoun, 2017; Weise et al., 2019) have argued that higher education should focus not only technical skills but also on uniquely human skills that are less likely to be rendered irrelevant by AI. For instance, Joseph Aoun (2017) has proposed a model of learning called "humanics" that includes three literacies (technological, data, and human); human literacy includes creativity, critical thinking, entrepreneurship, and the ability to navigate diverse cultures. Aoun has argued that these skills are best learned through experiential opportunities that allow students to use these skills in an integrated manner. Both Aoun and others have also argued about the need move away from the stereotypical view of higher education as a distinct phase that one passes through, typically in their late teens and early twenties, on the way to a subsequent career. Instead, for a number of reasons (e.g., technical expertise is likely to become outmoded more quickly, AI may displace some careers over time, people may have a longer life span), people should be prepared to change careers more often in the future; thus, returning for further education will become a more frequent event. Such lifelong learning episodes are likely to vary in length and depth of training (e.g., short-term training certificates versus graduate degrees), and although there are many models for such training, it would typically involve collaboration between educational entities and employers as well as other stakeholders (e.g., Weise et al., 2019). Like many of the developments discussed in this chapter, the nature of higher education in an AI future is only beginning to be sketched out, and Aoun's vision is one of many. However, given other forces for change, we can safely anticipate that the experience of students 20 years from now is likely to be quite different than it is today.

FINAL THOUGHTS

Cognitive science examines the basic processes of perception, attention, and decision making, abilities made possible by a human brain that evolved in relatively simple informational environments. But although our brain is adapted for these environments, we live in an era in which we have instant access to information and are surrounded by technologies that vie for our attention more strongly than ever before. In such circumstances, our attention is a valuable, but limited, resource (the attention economy; Davenport & Beck, 2001). The results of the psychological and neuroscientific research described in this volume suggest ways of more deeply understanding how human experience has been impacted in this new environment. In this chapter, we have discussed the challenges we face as the pace of technological change continues to quicken as well as potential avenues for improving the possible outcomes of this change.

Besides the broad "interventions" we have described as being key to a successful technological future, we also believe in the importance of new research focused at both basic and applied levels, and the previous chapters have articulated new directions for this work on a variety of topics. Furthermore, we emphasize that both types of research are critical. Testing the implications of basic theories of human cognition and emotion in complex real-world environments can point to new theoretical insights or gaps in knowledge, and, conversely, the results of laboratory-based research can inform the design of technologies that enhance human experience and performance (e.g., Lane & Meissner, 2008). We will need rigorous scientific research, conducted inside and outside the laboratory, to guide the many decisions ahead—especially the ones we have not yet anticipated.

All of the contributors to this volume have backgrounds in psychology, neuroscience, or both, and this volume is intended to highlight research from these disciplines that is relevant to understanding how our lives are being impacted by technology. But understanding and addressing the challenges humans face from advancing technology is fundamentally an interdisciplinary problem. For example, advances in vehicle safety are likely to come from collaborations of automotive engineers, traffic engineers, computer scientists, human factors professionals, bioengineers, and researchers, among others. Even more broadly, the types of challenges discussed in this chapter touch on nearly every discipline taught in a modern university: business, engineering, science, medicine, law, arts, humanities, social sciences, and education. Solving the challenges ahead will require not only collaboration but integration of knowledge across disciplines, and it will undoubtedly lead to the rise of entirely new fields. For instance, both the Massachusetts Institute of Technology (MIT News Office, 2018) and Stanford University (Adams, 2019) recently announced large-scale educational and research efforts examining issues surrounding the intersection of human and AI. The profound and complex nature of the problems to be solved means that this is an especially exciting time to be a student with interests in these issues or even a midcareer professional seeking a change.

Ultimately, we are neither blindly optimistic or pessimistic about how advancing IT will change the human experience. As with other periods of rapid technological change we have faced as a species (e.g., the industrial revolution; Stearns, 2012), our lives are being disrupted in both exciting and disconcerting ways. Improvements in AI and robotics will likely challenge us even more in the coming decades. We believe a positive outcome is unlikely if we do not ask deep questions about what it means to be human in a technological world, when and how technology should support that humanity, and how we can achieve that outcome. If we are willing to engage in this way, we have an opportunity to create a world in which technologies enhance our humanity rather than detract from it. We hope that the chapters in this volume have stimulated your own thinking and prompted you to ask questions about your life in the attention economy.

REFERENCES

Adams, A. (2019, March 18). Stanford University launches the Institute for Human-Centered Artificial Intelligence. *Stanford News.* https://news.stanford.edu/2019/03/18/stanford_university_launches_human-centered_ai/

Alexander, B., Adams Becker, S., & Cummins, M. (2016, October). *Digital literacy: An NMC Horizon Project strategic brief* (Vol. 3.3). The New Media Consortium. https://library.educause.edu/~/media/files/library/2016/6/2016stratbrief digitalliteracy.pdf

Alexander, B., Adams Becker, S., Cummins, M., & Hall Giesinger, C. (2017, August). *Digital literacy in higher education, Part II: An NMC Horizon Project strategic brief* (Vol. 3.4). The New Media Consortium. https://library.educause.edu/-/media/files/library/2017/8/2017nmcstrategicbriefdigitalliteracyheii.pdf

Allan, P. (2016, December 7). *Make your smartphone less distracting by switching your screen to grayscale.* Lifehacker. https://lifehacker.com/make-your-smartphone-less-distracting-by-switching-your-1789747192

Anders, G. (2017). *You can do anything: The surprising power of a "useless" liberal arts education.* Back Bay Books.

Anderson, C. A., Azron, C., Beach, M. A., Boninger, F., Sumerson, J. B., Brunelle, E., Buckley, W., Bushman, B. J., Cash, H., Cozen, J., Deak, J., Domoff, S. E., Fagen, D., Flores, L., Fortney-Parks, A., Fowers, B. J., Freed, R., Fundora, V., Gentile, D. A., . . . Young, K. (2018, August 8). *Our letter to the APA.* Children's Screen Time Action Network. https://screentimenetwork.org/apa

Anderson, J., & Rainie, L. (2018, April 17). *The future of well-being in a tech-saturated world.* Pew Research Center. http://www.pewinternet.org/2018/04/17/the-future-of-well-being-in-a-tech-saturated-world/

Angwin, J., Larson, J., Kirchner, L., & Mattu, S. (2017, April 5). *Minority neighborhoods pay higher car insurance premiums than White areas with the same risk.* ProPublica. https://www.propublica.org/article/minority-neighborhoods-higher-car-insurance-premiums-white-areas-same-risk

Angwin, J., Larson, J., Mattu, S., & Kirchner, L. (2016, May 23). *Machine bias.* ProPublica. https://www.propublica.org/article/machine-bias-risk-assessments-in-criminal-sentencing

Angwin, J., Scheiber, N., & Tobin, A. (2017, December 20). Facebook job ads raise concerns about age discrimination. *The New York Times.* https://www.nytimes.com/2017/12/20/business/facebook-job-ads.html

Aoun, J. E. (2017). *Robot-proof: Higher education in the age of artificial intelligence.* MIT Press.

Arntz, M., Gregory, T., & Zierahn, U. (2016). *The risk of automation for jobs in OECD countries: A comparative analysis* (OECD Social, Employment, and Migration Working Papers, No. 189). OECD Publishing.

Atchley, P., & Lane, S. (2014). Cognition in the attention economy. In B. H. Ross (Ed.), *The psychology of learning and motivation* (Vol. 61, pp. 133–177). Academic Press.

Athey, S. (2015, August). Machine learning and causal inference for policy evaluation. In C. Longbing & C. Zhang (Chairs), *Proceedings of the 21st ACM SIGKDD International Conference on Knowledge Discovery and Data Mining* (pp. 5–6). Association for Computing Machinery. https://doi.org/10.1145/2783258.2785466

Baer, T., & Kamalnath, V. (2017, November). *Controlling machine-learning algorithms and their biases*. McKinsey & Company. https://www.mckinsey.com/business-functions/risk/our-insights/controlling-machine-learning-algorithms-and-their-biases

Bellman, S., Johnson, E. J., Kobrin, S. J., & Lohse, J. L. (2004). International differences in information privacy concerns: A global survey of consumers. *Information Society, 20*(5), 313–324. https://doi.org/10.1080/01972240490507956

Bunch, E. (2018, January 23). *12 unplugged destinations for an epic, digital-detox vacay*. Well Good. https://www.wellandgood.com/good-travel/tech-free-unplugged-vacation-ideas-for-2018/

California Consumer Privacy Act, California Civil Code § 1798–100, 2018.

Carr, N. (2010). *The shallows*. W. W. Norton.

Center for Humane Technology. (n.d.-a). *Design guide (alpha version)*. https://humanetech.com/designguide/

Center for Humane Technology. (n.d.-b). *The way forward*. https://humanetech.com/problem#the-way-forward

Common Sense Education. (n.d.). *Everything you need to teach digital citizenship*. https://www.commonsense.org/education/digital-citizenship

Corbett-Davies, S., Pierson, E., Feller, A., & Goel, S. (2016, October 17). A computer program used for bail and sentencing decisions was labeled biased against Blacks. It's actually not that clear. *The Washington Post*. https://www.washingtonpost.com/news/monkey-cage/wp/2016/10/17/can-an-algorithm-be-racist-our-analysis-is-more-cautious-than-propublicas

Darrow, B. (2017, May 24). Automation, robots, and job losses could make universal income a reality. *Fortune*. http://fortune.com/2017/05/24/automation-job-loss-universal-income

Davenport, T. H., & Beck, J. C. (2001). *The attention economy: Understanding the new currency of business*. Harvard Business School Press.

Davidson, C. N. (2017). *The new education: How to revolutionize the university to prepare students for a world in flux*. Basic Books.

Deloitte. (2017). *2017 Global Mobile Consumer Survey: U.S. edition*. https://www2.deloitte.com/content/dam/Deloitte/us/Documents/technology-media-telecommunications/us-tmt-2017-global-mobile-consumer-survey-executive-summary.pdf

Edwards, H. S. (2018, April 13). You're addicted to your smartphone. This company thinks it can change that. *Time*. http://time.com/5237434/youre-addicted-to-your-smartphone-this-company-thinks-it-can-change-that/

Electronic Information or Data Privacy Act, H.R. 57, §23c-102(1)(a) (2019). https://le.utah.gov/~2019/bills/static/HB0057.html

Ent, M. R., Baumeister, R. F., & Tice, D. M. (2015). Trait self-control and the avoidance of temptation. *Personality and Individual Differences, 74*, 12–15. https://doi.org/10.1016/j.paid.2014.09.031

European Commission. (n.d.). *EU data protection rules*. https://ec.europa.eu/info/priorities/justice-and-fundamental-rights/data-protection/2018-reform-eu-data-protection-rules/eu-data-protection-rules_en

Fogg, B. J. (2002). *Persuasive technology: Using computers to change we think and do*. Morgan Kaufman.

Fogg, B. J. (2009, April). A behavior model for persuasive design. In S. Chatterjee & P. Dev (Chairs), *Persuasive '09: Proceedings of the 4th International Conference on Persuasive Technology* (Article No. 40, pp. 1–7). Association for Computing Machinery. https://doi.org/10.1145/1541948.1541999

Fortin, C. (2017, July 17). *Behind the scenes: A minimalist's smart phone.* New Minimalism Declutter & Design. http://www.newminimalism.com/blog/minimalist-phone

Frank, M. R., Autor, D., Bessen, J. E., Brynjolfsson, E., Cebrian, M., Deming, D. J., Feldman, M., Groh, M., Lobo, J., Moro, E., Wang, D., Youn, H., & Rahwan, I. (2019). Toward understanding the impact of artificial intelligence on labor. *Proceedings of the National Academy of Sciences, 116*(14), 6351–6359. https://doi.org/10.1073/pnas.1900949116

Frey, C. B., & Osborne, M. A. (2017). The future of employment: How susceptible are jobs to computerisation? *Technological Forecasting and Social Change, 114*, 254–280. https://doi.org/10.1016/j.techfore.2016.08.019

Granville, K. (2018, March 19). Facebook and Cambridge Analytica: What you need to know at fallout widens. *The New York Times.* https://www.nytimes.com/2018/03/19/technology/facebook-cambridge-analytica-explained.html

Guszcza, J., Rahwan, I., Bible, W., Cebrian, M., & Katyal, V. (2018). *Why we need to audit algorithms.* Harvard Business Review. http://hdl.handle.net/21.11116/0000-0003-1C9E-D

Handley, L. (2018, November 9). *The "world's first" A.I. news anchor has gone live in China.* CNBC. https://www.cnbc.com/2018/11/09/the-worlds-first-ai-news-anchor-has-gone-live-in-china.html

Kurzweil, R. C. (2006). *The singularity is near.* Penguin Books.

Lane, S. M., & Meissner, C. A. (2008). A "middle road" approach to bridging the basic-applied divide in eyewitness identification research. *Applied Cognitive Psychology, 22*(6), 779–787. https://doi.org/10.1002/acp.1482

Manyika, J., Lund, S., Chui, M., Bughin, J., Woetzel, J., Batra, P., Ko, R., & Sanghvi, S. (2017). *Jobs lost, jobs gained: Workforce transitions in a time of automation.* McKinsey Global Institute. https://www.mckinsey.com/featured-insights/future-of-work/jobs-lost-jobs-gained-what-the-future-of-work-will-mean-for-jobs-skills-and-wages

Mark, G., Czerwinski, M., & Iqbal, S. T. (2018, April). Effects of individual differences in blocking workplace distractions. In R. Mandryk & M. Hancock (Chairs), *CHI '18: Proceedings of the 2018 CHI Conference on Human Factors in Computing Systems* (Paper No. 92, pp. 1–12). Association for Computing Machinery. https://doi.org/10.1145/3173574.3173666

Mickle, T. (June 4, 2018). Apple unveils ways to help limit iPhone usage. *The Wall Street Journal.* https://www.wsj.com/articles/apple-unveils-ways-to-help-limit-iphone-usage-1528138570

MIT News Office. (2018, October 15). MIT reshapes itself to shape the future. *MIT News.* http://news.mit.edu/2018/mit-reshapes-itself-stephen-schwarzman-college-of-computing-1015

Nedelkoska, L., & Quintini, G. (2018). *Automation, skills use and training* (OECD Social, Employment and Migration Working Papers No. 202). OECD Publishing. https://doi.org/10.1787/2e2f4eea-en

Nietzel, M. T. (2019, January 7). Whither the humanities: The ten-year trend in college majors. *Forbes.* https://www.forbes.com/sites/michaeltnietzel/2019/01/07/whither-the-humanities-the-ten-year-trend-in-college-majors

Organisation for Economic Co-operation and Development. (2013). *OECD skills outlook: First results from the Survey of Adult Skills*. OECD Publishing.

Osoba, O. A., & Welser, I. V. W. (2017). *An intelligence in our image: The risks of bias and errors in artificial intelligence*. RAND Corporation. https://www.rand.org/pubs/research_reports/RR1744.html https://doi.org/10.7249/RR1744

Pangburn, D. J. (2019, February 8). Washington could be the first state to rein in automated decision-making. *Fast Company*. https://www.fastcompany.com/90302465/washington-introduces-landmark-algorithmic-accountability-laws

Press Office, City of New York. (2018, May 16). *Mayor de Blasio announces first-in-nation task force to examine automated decisions used by the city*. Official Website of the City of New York. https://www1.nyc.gov/office-of-the-mayor/news/251-18/mayor-de-blasio-first-in-nation-task-force-examine-automated-decision-systems-used-by

Rainie, L., & Anderson, J. (2017, May 3). *The future of jobs and jobs training*. Pew Research Center. http://www.pewinternet.org/2017/05/03/the-future-of-jobs-and-jobs-training/

Schilhab, T. S. S., Stevenson, M. P., & Bentsen, P. (2018). Contrasting screen-time and green-time: A case for using smart technology and nature to optimize learning processes. *Frontiers in Psychology, 9*, 1–5. https://doi.org/10.3389/fpsyg.2018.00773

Selingo, J. J. (2013). *College (un)bound: The future of higher education and what it means for students*. Amazon.

Shlain, T. (2013, March 1). Technology's best feature: The off switch. *Harvard Business Review*. https://hbr.org/2013/03/techs-best-feature-the-off-swi

Simon, H. A. (1971). Designing organizations for an information-rich world. In M. Greenberger (Ed.), *Computers, communication, and the public interest* (pp. 37–52). The Johns Hopkins University Press.

Singer, N. (2018, July 26). Amazon's facial recognition wrongly identifies 28 lawmakers, A.C.L.U. says. *The New York Times*. https://www.nytimes.com/2018/07/26/technology/amazon-aclu-facial-recognition-congress.html

Stearns, P. N. (2012). *The industrial revolution in world history* (4th ed.). Routledge.

Stolzoff, S. (2018, February 1). The formula for phone addiction might double as a cure. *Wired Magazine*. https://www.wired.com/story/phone-addiction-formula/

Turkle, S. (2011). *Alone together*. Basic Books.

Weise, M. R., Hanson, A. R., Sentz, R., & Saleh, Y. (2019). *Robot-ready: Human+ skills for the future of work*. Strada Institute for the Future of Work. https://www.economicmodeling.com/robot-ready-reports/

Whittaker, Z. (2018, November 30). *Marriott says 500 million Starwood guest records stolen in massive data breach*. TechCrunch. https://techcrunch.com/2018/11/30/starwood-hotels-says-500-million-guest-records-stolen-in-massive-data-breach/

Wilson, H. J., & Daugherty, P. R. (2018). *Man+machine: Reimagining work in the age of AI*. Harvard Business Review Press.

Index

About the Editors

Sean Lane, PhD, is a professor of psychology and dean of the College of Arts, Humanities, and Social Sciences at the University of Alabama in Huntsville. Dr. Lane's research examines the mechanisms underlying learning and memory, and how these mechanisms influence behavior in complex real-world settings. He previously worked in the technology industry in the area of user experience and human factors. Dr. Lane has worked to further the productive interaction between basic and applied research, including serving on the governing board of the Society for Applied Research in Memory and Cognition and as associate editor of *Applied Cognitive Psychology*. He received his PhD from Kent State University.

Paul Atchley, PhD, is currently on faculty at the University of South Florida, Tampa. Dr. Atchley has been conducting research and teaching about cognitive factors related to driving for more than 25 years. He received his PhD from the University of California, Riverside, in 1996 and completed postdoctoral training at the Beckman Institute for Advanced Science & Technology at the University of Illinois in 1998. He has published numerous peer-reviewed articles and chapters on issues of vision and attention, including their relationship to driving. He also has received awards for his service, research, teaching, and student advising. Dr. Atchley's work has been highlighted by the national and international press, such as the BBC, NPR, *TIME*, and *The New York Times*. He is part of efforts at the state and national levels to reduce distracted driving.